THE SEA OF ENERGY

T. Henry Moray

pictured as in office

THE SEA OF ENERGY
from the original

THE SEA OF ENERGY IN WHICH THE EARTH FLOATS

by T. Henry Moray

Revised and Reprinted
1978 from the 1960
4th edition

History and Biography by
John E. Moray

First Printing 1930
as "Beyond the Light Rays"
copyrighted 1945 and 1956 by
T. Henry Moray

Printed in the U.S.A.

5th edition

Published by
Cosray Research Institute, Inc.
1978
copyrighted

IN MEMORY OF MY FATHER

TABLE OF CONTENTS

LIST OF ENCLOSURES

CHAPTER 6

T. Henry Moray, 1932

INTRODUCTION TO 4th EDITION

RADIANT ENERGY - the term Moray has used to describe that source of energy coming from the cosmos to the earth and radiating from the earth back from whence it came. This is the energy the Moray device captures and could be described as those particles of energy pervading all space. In the evolution of energy and the evolution of matter these particles of matter and energy (one and the same) manifest under certain conditions as pure energy and under others as pure matter. Radiant Energy from the cosmos, like radiant particles of matter, being composed of an infinitesimal quantity whose behaviors are described by mathematical equations similar to those used for describing electrical waves, keeping in mind to differentiate between wave length and frequency. Radiant Energy is particles of energy, just as light is wavelengths and particles are comparable to the electron and magneton: a ring of negative electricity traveling in a vortex with the speed of light, streams of energy quanta, each quantum having energy and momentum where the electron revolves around the proton at a distance equal to the electron radius.

To Summarize — Radiant Energy as herein used is that energy existing in the luminiferous medium of the universe, kinetic and exercised in wave transmission and rendered sensible by conversion of its energy into a detectable frequency. IN THE FINAL ANALYSIS RADIANT ENERGY IS A MEANS OF USING THE ENERGY RELEASED BY THE FISSIONABLE REACTIONS TAKING PLACE IN THE STELLAR CRUCIBLES OF THE UNIVERSE.

INTRODUCTION TO THE 5th EDITION

At a recent Kiwanis meeting I heard a lecture given by the public relations officer of the Bell Telephone Company. The speaker illustrated his lecture with a historical film by Bell Laboratories justifying their position as a research organization and a leader in communications. It was interesting to note that although the film was essentially correct, the story told did not coincide in all particulars with what actually took place. For example, Bell gave Mr. Edison the credit for the electric light. He definitely deserves a great deal of credit. However, Edison only took a known principle of the day and produced a workable lamp globe by discovering through tedious trial and error the one thing, the tungsten filament, that would stand up under the crude vacuum systems then available to him. Naturally enough, the Bell film also gave Alexander Graham Bell great credit for the telephone. But it was really Joshius Coppersmit who first made a practical demonstration of a telephone. Likewise, the narrator in the film spoke of how Mr. Farnsworth had developed television. In fact, the first television demonstration was made by Francis Jenkins to Secretary of the Navy Wilber at the old Naval Radio Station in Washington, D.C., in June 1925.*

Research organizations such as Bell Labs have succeeded in large part because they have been able to inform the world of their work and thus market it. On the other hand, many young men — Francis Jenkins, for example — have come and gone, disappearing into obscurity, making great discoveries but receiving no credit for them, unable to develop them fully. I think one of the greatest such discoveries was the invention of radio by the Spanish physicist, Slava. One hundred and fifty years before radio was generally known, Slava transmitted radio signals from an island off the coast of Spain back to the mainland.

Fifty years before the advent of the atomic bomb, Dr. Gustav LeBon published a book describing what a nuclear fission reaction would be like, estimating it would be equivalent to the energy released by "1,340,000 barrels of gun powder." I am told that, years later, Albert Einstein wrote a letter to LeBon's daughter stating that Dr. LeBon, physician turned physicist, was truly the father of nuclear physics. Though generally unheralded, LeBon's position as a nuclear pioneer can easily be verified simply by going to one of our better stocked public libraries and checking out his book, **The Evolution of Matter.**

At this point the reader may well ask what my concern is with such obscure

**CBS, "Industries of Tomorrow," April 24, 1937*

individual research. The reader may also question my interest in scientific experiments whose results do not behave in accord with accepted scientific dogma. My answer is that I am a scientist and the son of a scientist, and I desire to see facts published as they really are. I also wish to present some unorthodox theories.

In the first edition of my father T. Henry Moray's book, **The Sea of Energy in Which the Earth Floats,** the history of the Moray radiant energy device was secondary, contained only in one chapter near the end. I wish, however, to emphasize that the workability of this device was repeatedly demonstrated, and it was successfully reproduced many times by Dr. Moray. Consequently, in this extensive revision of my father's book, I have introduced the subject of Radiant Energy with the history, leaving the theory to the last.

If, in repeating what my father felt and thought in the past, I should offend some innocent bystander or "well-meaning" individuals who did not intend offense to Dr. Moray, then all I can say is that the purposes of science sometimes are inadvertently obstructed by well-meaning individuals, and I will be the first to apologize if someone will point out my error. The historical information is as I have received the notes and as I remember it, living with my father some 45 years.

In this book I have gone into the historical background to try to interpret Dr. Moray's feelings. In writing his history I have taken the liberty to give some of my own opinions since I lived through many of these events and they are encircled with emotion for me.

Dr. Moray's work, unfinished as it is, can stand the strain of criticism. And, hopefully, its publication at this time will excite the interest and support it deserved originally. At times during the story's unfolding it becomes impossible for me to eliminate entirely my father's religious convictions. However, I wish to emphasize that at no time did he allow religion to become involved in his work, lest any man should say that he used religion for his own aggrandizement, and therefore imply that he should be found lacking before the Lord in whom he so fervently believed.

Although he was frequently misunderstood, at all times he exhibited rectitude. Only once did I see him "lose his cool," and that was when asked to comment on the "dirty tricks" department of life.

The story of Dr. T. Henry Moray is the story of a man working alone, whose discoveries and theories were years, even decades, ahead of his time. His many pamphlets and writings advanced ideas with which science is only now coming more and more into agreement. His active mind delved into many fields of science and always came up with new and startling discoveries.

Time and space do not allow a full discussion of the various Moray devices, since the primary subject of this book is a presentation of Radiant Energy (RE). A very complete and detailed record of letters, articles, and pamphlets are on file in the records of The House of Moray in Salt Lake City, Utah.

Dr. Moray and Radiant Energy are so closely related that one cannot be separated from the other. It is not the purpose of this history to give a complete picture of Radiant Energy. But a discussion of it is necessary since all of the other work with which Dr. Moray was involved seems to be received on the basis of how the party or parties involved feel about Radiant Energy. Since 1926 he paralleled his work on Radiant Energy with study in a number of other fields, the most important among them being his intensive studies of radiation and radioactive substances.

Through the years, Dr. Moray's chief wish was to be free of bothersome business details so that he could concentrate on research. Yet every time he turned business details over to someone else, chaos resulted. One of his chief difficulties was to find the unusual person who combined the talents of a good business manager with a knowledge of science and who was dedicated, truthworthy, unselfish, and tough enough to withstand the rough road which had to be traveled. The qualities of truthworthiness and toughness seem to have been lacking most often. One must admit that Dr. Moray did an outstanding job in keeping his private Utah corporation alive in the entity of Research Institute Incorporated. His personal business ability was above question. The only fault to be found (if one can call it a fault) is that he was too willing to forgive those with selfish interests who attempted to destroy his work.

Throughout history, progress has been slowed and at some points stopped by selfish interests. But in spite of all obstacles, mankind crawls slowly forward. Similarly, the work of T. Henry Moray has been slowed to a crawl; nonetheless, it painstakingly continues.

On his deathbed, my father said to me, "I only wanted to finish it! If only I had been given a chance. No one even went half-way with me."

But our father taught my brother and me his dream. We are endeavoring to carry it out. An "impossible dream"? Perhaps, but we believe that we can make it come true — and that our world will be the better for it.

Dr. T. Henry Moray in Salt Lake City, Utah October, 1970

FOREWORD

When John Moray asked me to assist with the rewrite and editing of this new edition of **The Sea of Energy in Which the Earth Floats,** I was both delighted and deeply appreciative. For several years I have devoted a great amount of time in the research for the theory — even an **ad hoc** theory — that could possibly explain the profound achievement of John's father, Dr. T. Henry Moray, in his discovery and development of free radiant energy. And I can assure the reader that only through such a diligent and unrelenting search for an explanation can the magnitude of T. Henry Moray's accomplishment truly be appreciated, or the breadth of his vision be grasped. Even today, decades after Moray built his successful free energy device, the most advanced physics theory is not quite capable of encompassing his work, though it has approached it ever more closely as the years have passed. If comprehension were not formidably difficult, rediscovery of the device would have occurred over and over, and free energy would already be available to every household. Though Moray was a relatively simple man, nationally unknown, he was an engineer of excellent ability and strong character. He was a giant, an experimental genius who was far ahead of his own time.

The story of Dr. T. Henry Moray and his free energy device is incredible but true. There really are two stories involved. One is the story of a courageous and lone pioneer who attempted the impossible and succeeded, who wrested from nature one of her most zealously guarded secrets to produce a device which could literally have ushered in a golden age for all mankind. The second is the story of the greed, arrogance, hostility, and stony disbelief that greeted all his efforts to introduce his great invention into the service of mankind.

Make no mistake about it, Dr. Moray did what he claimed to have done. He achieved free energy. He also built the forerunner of the transistor long before such a device was dreamed of by Western scientists. If people had listened and his work had been recognized, there would be no energy crisis and no energy problem today. Everyone could have energy sufficient for his needs, and the vast amounts of unburned hydrocarbons that pour into our atmosphere today from the burning of fossil fuels would be reduced to a mere trickle. Cheap, portable, plentiful power would be available to every underdeveloped country in the world. This was T. Henry Moray's dream — a dream that was thwarted through no fault of his own, in spite of his prodigious efforts to overcome the indifference and cynicism which continually faced him.

Moray demonstrated his device over and over — even allowing prominent scientists to visit his laboratory, disassemble his equipment and satisfy themselves that no trickery or fraud was involved, then run the experiment themselves to produce free energy. Yet, try as he would, he was unable to overcome the ignorance and bias which greeted him on every turn. The scientists of his day would not believe the demonstration even when they themselves conducted it, and the patent office would not grant him a patent.

I am particularly chagrined because, while our Western scientists were castigating Moray's work at every turn and successfully suppressing it, more dictatorial societies were earnestly seeking to obtain Moray's services and his invention. Imperial Japan sought to bring Moray to Japan and have him build a death ray — for Moray had shown that a beam of radiant energy of sufficient intensity could simply destroy every living thing in its path. The Soviet Union offered to give him his own fully equipped laboratory in Russia, with no expense spared, and to back his experiments fully. One man — who in my personal opinion was a trained Soviet agent — even succeeded in working his way into Moray's confidence and gaining access to Moray's laboratory as a technician and assistant. When Moray still refused to give his invention and services to the Soviet Union, the assistant destroyed the device, smashing it to pieces with a hammer.

Shortly thereafter Moray was assaulted and shot in his own laboratory. Except for his own skill with a pistol to successfully defend himself against his assailants, Moray would have been murdered. Repeated assassination attempts were made against his life; it was necessary for him to bulletproof his automobile since he was shot at while driving down the public street. Small wonder that Moray developed an extremely alert and suspicious nature, and visitors to his desk often noticed a fully loaded pistol lying on the desktop within easy reach of his hand! Sadly, T. Henry Moray died with his dream unrealized and the original device destroyed.

But the House of Moray lives on. John and Richard Moray have endeavored to continue their father's work and to bring their father's dream to the fruition it so richly deserves. Yet the same hostility and derision have greeted them at almost every turn. However, the Morays are made of stern stuff, and John and Richard have continued to the best of their ability and resources. I have found John a man of great courage, strong determination, complete sincerity, and down-to-earth wisdom. I consider it an honor to be his friend. I am also determinded to contribute whatever I can to an understanding of the Moray radiant energy device and to the realization of T. Henry Moray's dream.

In my opinion, time is of the essence in telling the true story of Dr. T. Henry Moray's achievement. We see an economic noose being slowly drawn around the neck of the world by the oil-producing countries. They have the power to destroy civilization economically whenever they wish, simply by cutting off the oil flow. The raw impact of this power is already being felt in world diplomacy, and it is not at all exaggerated to say that the "big stick" the U.S. used to carry has shrunk considerably of late because of the problem of energy. Oil blackmail is already a potent force on the international scene, and if we are to avert economic disaster, we must find a cheap new source of bountiful energy quickly.

Today we can state unequivocally that such a source is present everywhere in the universe, free for the taking if we can become knowledgeable enough. At the present state of development of quantum mechanics, quantum electrodynamics, and geometrodynamics, we know that each tiny portion of space — pure empty vacuum — contains almost infinite energy. According to Wheeler's calculation, the available energy in one cubic centimeter of pure vacuum is greater than 10^{100} grams (expressed in mass units), an incredible packet of raw energy sufficient to provide for the formation of entire solar systems. Zero-point energy of the vacuum is essentially infinite and free for the taking, if we are clever enough to discover how to do it. Further, we know that this energy can be tapped because the lowly hydrogen atom already does it — the Lamb shift is in fact due to the "tapping" of a tiny bit of this vast storehouse of power. And that is precisely what Dr. T. Henry Moray did, long before zero-point energy as such was known, and long before Wheeler made his calculation.

It is only because of some of the most theoretically advanced concepts of today that some of the simple but profound concepts of Dr. T. Henry Moray can be understood. Let us hope that with the advent of the new edition of this book, his dream of plentiful power for mankind can yet be realized.

Tom Bearden

June 4, 1977

CHAPTER 1
EARLY YEARS

"Enough energy is coming to the earth to light over 1.5 million (1,693,600) 100-watt lamps for every human being on the earth today. No fuel of any kind need be taken as a dead load since this energy can be "picked-up" directly by ocean liners, railroads, airplanes, automobiles, or any form of transportation. Heat, light and power can be made available for use in all kinds of buildings and for all kinds of machinery. An example would be to pump water onto the desert lands, the power source being only a fraction of the weight of any steam plant or any kind of engine in use today, and all this at a fraction of the current cost.

"A wild dream? No! It's a proven practical reality, as hundreds of people know who have witnessed the Moray radiant energy invention — powered from the cosmos." So stated Thomas Henry Moray's original **Sea of Energy In Which the Earth Floats.**

In order to understand the development of Radiant Energy, one must understand what took place in Dr. Moray's life. By looking at the background of the Moray family and of the family of his mother, Petronella Larson, one can understand why Dr. Moray was in a sense a recluse and refused in some instances to discuss his invention with individuals who might otherwise have given him great credit.

Henry Moray was the product of emigrants, a Swedish mother and Irish father. Petronella Larson came from Sweden during a period when superstition was very great and under circumstances that required her to sacrifice greatly in order to immigrate from Sweden through Pennsylvania to Salt Lake City. She married James Cain Moray, who had been born in Ireland to a family living in hiding because its members had been condemned since the days of Bonnie Prince Charlie and the Scottish-English wars. This marriage was filled with disaster: of seven children, only two survived to grow to maturity - Henry Moray and his sister, Nellie.

A family fortune built by Henry's father, James Cain Moray, was almost entirely lost after the father died, and individuals who supposedly could be trusted betrayed Henry's mother. As a result, Henry's mother wished him to follow a business career and encouraged him by offering to buy up businesses with what little money was left. She insisted that he attend the old LDS College because it had a business course.

Henry Moray's interests leaned toward electronics and electrical engineering rather than business. His work was confined to whatever he could "scrounge" up himself as his mother thought his interests a waste of time. He often would go to the garbage dump to find pieces of wire, scraps of tape, and material he wanted to work with in the basement. At one time his uncle, John Moray, offered to put up money for the young man's experimenting, but Henry's mother became too upset.

At age fifteen he obtained a job as an electrician, wiring houses. It would seem that a large number of the houses being wired at the time happened to be houses of prostitution. I heard him mention on many occasions that the girls always treated him with great respect and the madame made them behave themselves when he was there working. He felt particularly sorry for one of the girls and tried to show her the error of her ways. In later years, subsequent to his mission for the L.D.S. Church, he visited a business acquaintance and was pleasantly surprised to find that the girl was now married to the businessman.

His thoughts were continually drawn to an idea pounding in his brain and looking for expression. In a brief history of the Radiant Energy device, he wrote:

"I started my experimentation with the taking of electricity from the ground, as I termed it, during the summer of 1909. By the fall of 1910 I had sufficient power to operate a small electrical device, and I made a demonstration of my idea to two friends, Lorine and Parnel Hinckley. This demonstration in the early stages of my experimenting consisted of operating a miniature arc light. I had the idea of using what I termed static, based upon the experiment of Ben Franklin with his kite, but as advancements were made it soon became evident that the energy was not static and that the static of the universe would be of no assistance to me in obtaining the power I was seeking.

"Although the accepted scientific ideas of the day seemed to point to the impossibility of what I was trying to do, there was never a time when, within my mind, anything but ultimate success was assured. Even when only enough energy was obtained to make a slight click in a telephone receiver, I did not doubt the soundness of the idea upon which I was working.

"During the Christmas Holidays of 1911, I began to fully realize that the energy I was working with was not of a static nature, but of an oscillating nature. Further I realized that the energy was not coming out of the earth, but instead it was coming to the earth from some outside source. These electrical oscillations in the form of waves were not simple oscillations, but were surgings - like the waves of the sea - coming to the earth continually, more in the daytime than at night, but always coming in vibrations from the reservoir of colossal energy out there in space. By this time I was able to obtain enough power to light the old type 16-candlepower carbon lamp for about one half capacity, and I did not seem to make any further improvement until the spring of 1925."

In 1910 or 1911 he enrolled in a correspondence course in electrical engineering in Pennsylvania.

In July of 1911, Petronella Larson bought for her son what they thought was the controlling interest in what was then known as the Independent Electric Company. They bought the company on the basis of a financial statement furnished them by certain prominent individuals in Salt Lake, depending entirely on the integrity of these individuals for the accuracy of the reports. It turned out that the inventory and accounts receivable had been falsified, and that most of the accounts receivable had been outlawed because no attempt to collect them had been made in a reasonable amount of time to keep them alive.

Henry went to work for the company, became secretary and a member of the board of directors, and became very friendly with the president, a man by the name of Holly, the biggest owner of the company besides Henry. Holly immediately understood that Henry had been taken advantage of. Together they got rid of the incumbent manager and attempted to salvage what they could of the company. It was three weeks before either of them fully realized the extent of the misrepresentation. Much to Henry Moray's surprise, he had bought only an interest in the inventory and the sales and had no control over management. The only thing that was correct was "bills payable." Even though Holly attempted to salvage the company (after Henry Moray was called on an L.D.S. Mission), Henry and his mother lost everything that they had put into the company. Only one of the individuals responsible ever tried to explain his own actions, and then he simply said that if he had known to whom he had sold the company, he would not have done so. Because this individual himself had lost so much money on the company, he thought he was justified in making the deal.

Years later R.L. Judd, an attorney and friend from Henry's LDS College days, took up the case and tried to straighten it out. But because of political pressure, Judd finally advised Henry that he should withdraw his legal action. Henry felt he could not fight the responsible individuals alone. No other attorney in Salt Lake would take the case, so he abandoned any effort to obtain retribution for the financial loss he had suffered.

In 1912 he was called to go on a mission for the Church of Jesus Christ of Latter-day Saints, and under a visitor's visa was allowed to enter Sweden during the Exhibition of 1912 in Stockholm. With permission of his mission president, he attended the University of Upsalla for short periods of time, taking courses by examination. He mentioned many times that he feared he would be caught by a certain Lutheran pastor and discovered to be a missionary for the Church of Jesus Christ of Latter-day Saints. This Pastor Aslive had become an arch rival since Henry Moray had succeeded in embarrassing him in several debates. The pastor, therefore, became enraged any time he and Henry Moray chanced to meet. Since Upsalla was a state school, Henry was forced to attend under the name of James Cain Larson, assuming his mother's maiden name and using the address of his mother's half-sister, Fru Cecelia Nelson (c/o Andrus Solomon Larson, Perstorp, Skone).

In his notebook, dated November 1, 1913, Henry included a memo that he had obtained material from a railroad car at Abisco, Sweden the previous summer, and material from the side of a hill. He made electric tests of these materials, taking them home to try each as a detector for his energy machine. Tests indicated that this soft, white stone-like substance might make a good "valve-like detector."* Some excerpts from the notebook follow:

* *This "valve-like detector" is what led Henry to do research in semi-conductive*

"November 13, 1913, time early this morning to test the stone again. Hard to buy wire and stuff to make coils. Used cardboard tube to work on. Batteries of no help in priming. Got hard rubber Widstens from Wilsons and bought some silk and pure wool cloth to use as static generator. Got some red sealing wax and tried to make vacuum tube, but no luck. Silver wire used on stone makes a rectifier.

"January 21, 1914, been alone a few days down at south end of Gefle Branch where I came to study. Took my gadget with me when I left Gefle Sunday evening, January 19. Guess time will not permit me to do much with gadget. Yet will do all I can. I know it will work.

"Tuesday, May 5, 1914, at south end, had fun at class today. (Ed: referring to the University of Upsalla.) Swedes make fun of my Skone Swedish, but I do them in a way they cannot understand English. Got four hours in the shop and lab, with a friend, without making explanation. He asked 'what that dumb radio set was, can't get anywhere much hearing that device.' Soon will be home, in about four months."

Before leaving Sweden, July 31, 1914, he had succeeded in completing his doctoral thesis, which advanced the idea that there was energy throughout space. It is unfortunate this writing is not available to us. His oral examinations were completed, although no commencement ceremony or certificate of graduation was issued at that time. Through the years he kept abreast of the latest developments in science through diversified reading. He built up a personal scientific library of extensive proportions.

On November 28, 1917 (Fig. 4), he married Ella Ryser. They had five children: Henry, Jr., Ella Evelyn, Sylvia, John Eugene, and Richard Ryser.

At the end of World War I it was difficult to get engineering jobs. Moray was forced to work as a fireman on one of the railroads. Later he was able to obtain work as a civil engineer for the Denver and Rio Grande Railroad. However, he detested working as a civil engineer. The drawing (see Fig. 5), which he made, illustrates how he felt. The engineers working on both sides of him smoked cigars. He used to say he knew that if he had not quit the job, he would have ended up smoking a pipe simply for self defense.

Still later he held various other positions, serving as an electrical engineer and designer for the Phoenix Construction Company (a sub-contractor for the Utah Power and Light Company), an assistant chief electrical engineer for the Aarastad Construction Company, and an assistant division electrical engineer for the Mountain States Telephone and Telegraph Company.

During this time, he designed the electrical layouts for many large buildings in Salt Lake City and throughout the West. He was the design engineer for what was known as Terminal, located west of Salt Lake City, which was the largest oil-cooled electrical switch yard in the world. He was a certified member of the American Association of Engineers (figure 6) and secretary of the Independent Electric Company.

(continued from p. 21)

materials, and from this soft white stone he developed his first Moray valve and the Moray valve that was used in some of the early Radiant Energy devices.

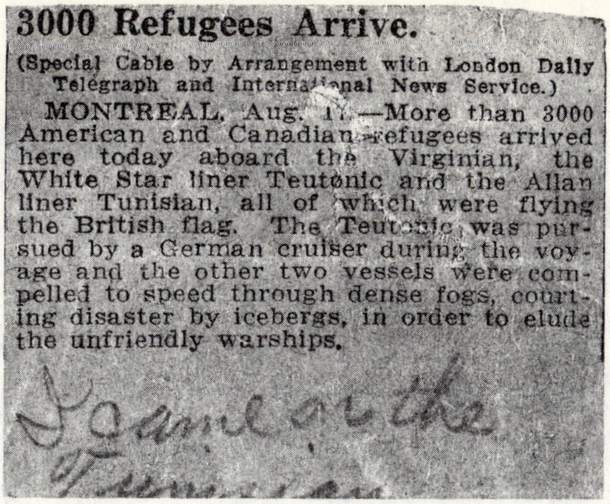

3000 Refugees Arrive.

(Special Cable by Arrangement with London Daily Telegraph and International News Service.)

MONTREAL, Aug. 1[illegible].—More than 3000 American and Canadian refugees arrived here today aboard the Virginian, the White Star liner Teutonic and the Allan liner Tunisian, all of which were flying the British flag. The Teutonic was pursued by a German cruiser during the voyage and the other two vessels were compelled to speed through dense fogs, courting disaster by icebergs, in order to elude the unfriendly warships.

I came on the

Fig. 3

Memo in newspaper, saved by Dr. Moray

Fig. 4

Nellie Moray Todd, top left, Chester M. Todd, back center, Petronella Larson, top right, James D. Todd, front left, and Ella Ryser Moray, front right.

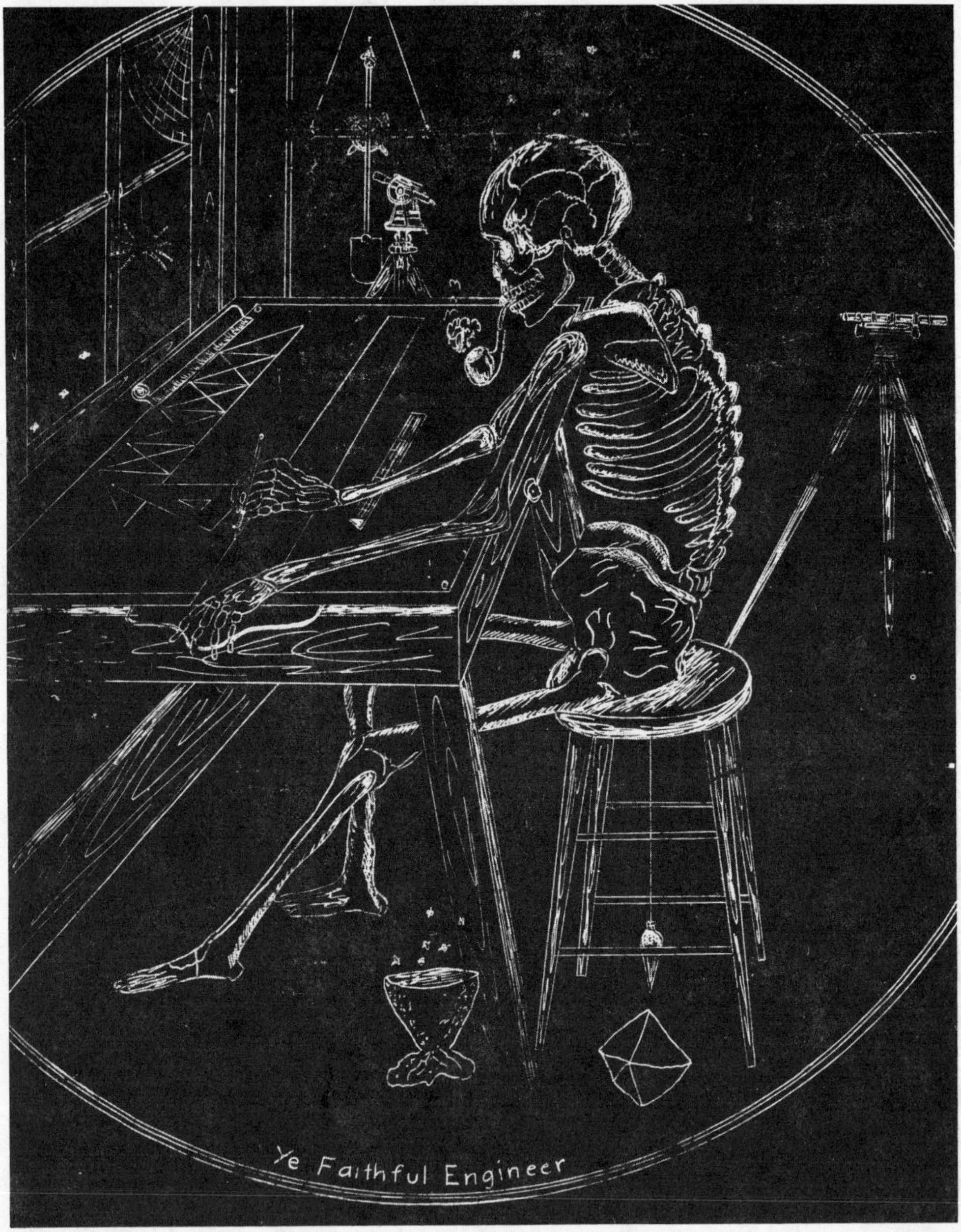

Fig. 5

Dr. T. Henry Moray's concept of Civil Engineering

On December 21, 1920, he was injured in an accident while examining a Utah Power & Light Company substation to determine the levels of inductive interference to telephone circuits. During this period the first attempts were made to discredit him. To obtain some compensation for both the labor time lost and the injury to his eyes, he was forced to take his case against the company all the way to the Utah Supreme Court. The transcript of the case reveals that the company had examined his background and everything about him.

He often mentioned that this accident may have been a blessing in disguise. His limited ability to see detailed work forced him from the drawing table into research and led him back into Radiant Energy. From time to time arguments were given that he had lost up to 75 percent of his vision. It is interesting to note that in the early 1970's, a few years before he died, an optometrist commented on the large number of burned spots of the retina of his eye. It was reported a few years ago that the Mountain Bell System denied that Henry had ever performed the work that resulted in his injury or that he had ever been an employee in their system. However, the Utah Supreme Court record exists today. A copy of the record is held in our files, as is his notebook on inductive interference.

The period that followed 1921 included some rather lean years for Henry Moray. He tried to expand into various other fields in order to occupy himself within the limitations of his eyesight. Without going into detail, it can be said that his financial woes multiplied until he was barely able to retain his home and that of his mother.

He obtained a certificate as a poultry judge and in September of 1923 he became a certified member of the American Poultry Association. He purchased 3000 Buff Orpington chicks and entered the chicken business. He had built a 150 foot long chicken coop behind his home. Properly housed and cared for, the Buff Orpington chicks grew into a fine flock in a short time. Henry then became interested in raising fine Cornish hens and purchased a hen, a cockerel, and some eggs. Before long these eggs hatched and the resulting chicks were raised and groomed into fine showbirds. He began winning prizes almost immediately and soon became known throughout the Salt Lake Valley as a raiser of spendid Cornish poultry.

So valuable was Moray's Cornish poultry that Dr. Warburton, a well-known Salt Lake dentist, once bought ten eggs from Henry at $50 each. Henry gave Dr. Warburton five extra eggs without charge and further guaranteed that these eggs would produce a prize-winning showbird. Not only that, he guaranteed that one of the birds would win Grand Champion. And sure enough, some time later one of the birds did win Grand Champion at the Utah Poultry Association Exhibit.

Henry's Cornish poultry soon became known throughout the state and beyond. He was asked to judge poultry at state fairs and at shows of the American Poultry Association in such faraway places as Montana, Kentucky, California, Kansas, Nebraska, and Vermont.

With such a widespread reputation as an expert raiser of Cornish poultry, it wasn't long before people were calling him "The Cornish King." Word of his Cornish poultry reached Hollywood, and producer Raoul Walsh and actress

Representatives in All Cities. Chapters and Clubs in All the Principal Cities

ADDRESS ALL COMMUNICATIONS TO

AMERICAN ASSOCIATION OF ENGINEERS
NATIONAL HEADQUARTERS
29 SOUTH LASALLE STREET
CHICAGO, ILLINOIS

ADVANCE · CO-OPERATE · AAE

TELEPHONE CENTRAL 73
CABLE ADDRESS "MONAD"

MEMBER OF CHAMBER OF COMMERCE OF THE UNITED STATES OF AMERICA

OFFICERS
F. H. NEWELL, PRESIDENT
W. W. DEBERARD, FIRST VICE-PRES.
T. A. EVANS, SECOND VICE-PRES.
C. E. DRAYER, SECRETARY
A. H. KROM, ASST. SECRETARY
F. H. MYERS, ASST. SECRETARY
JOHN ERICSON, TREASURER

July 19,
1 9 1 9

Mr. T. H/ Moray,
213 W. 4th North St.,
Salt Lake City, Utah,

Dear Sir:

~~Please be advised that you have been elected as a Certified Member~~ in this Association. Your membership year begins 6/15/19 and the annual membership dues are ten dollars. As soon as we receive your check for this amount you will be issued a membership card.

Each member is entitled to wear the emblem of the Association, a gold pin or button, which can be procured from this office for one dollar. A leather watch fob with a gold emblem is sold for two dollars.

The 1920 edition of the ENGINEERS' DIRECTORY will go to the printers July 31, 1919. The time is very limited to complete your membership and be classified. Please answer soon.

The enclosed membership Committee application blanks are for your engineering friends. Endorse each blank with your signature, as you will be given personal credit for those who join during the 1919 membership campaign. The Membership Committee asks each member to secure five new members during the year.

Yours for Co-operation,

C. E. Drayer

AHK/E

Secretary.

Money ($11) mailed July 22, 1919

OBJECTS TO RAISE THE STANDARD OF ETHICS OF THE ENGINEERING PROFESSION AND TO PROMOTE THE ECONOMIC AND SOCIAL WELFARE OF ENGINEERS

Fig. 6

American Association of Engineers

Thelma Todd visited him and bought many fine birds.

One particular Cornish cockerel was named Lord Cecil. Lord Cecil was a perfect bird (see figure 7), perfect in body features, color, and feathers. Whenever he was put on a table and viewed by judges, the opinions of the judges was unanimous. He always won Grand Champion. At the American Poultry Association's show at Sugarhouse, Salt Lake City, in 1923, Lord Cecil was judged by Judge Branch to be first cockerel. Judge Branch was a particularly difficult judge to please, and he commanded a fee of $500 a show for the job of judging. At that same show, Lord Cecil won the sweepstakes, the prize for the best color, and the prize for the best shape. After the show, he won the Grand Champion and the Gold Medal. The value of the bird at this time was at least $1,000.

Shortly before selling the major part of his breeding stock in Cornish Chickens to the movie actress, Thelma Todd, Henry Moray had to take legal action against one of the local chicken farmers who falsely claimed he had purchased Lord Cecil from Henry. Henry continued to sell prize eggs at a handsome price "by the setting," as it was called, for purchasers to raise chickens or show birds directly related to Lord Cecil. Henry then sold his breeding stock, planning to start over from the chicks he had left that year. Instead, he was wiped out completely. A marauding skunk or weasel got into the coop and destroyed his best cockerels.

In the early 1930's, Dr. Moray bought a prize horse by the name of Valencia Don for the sum of $5,000 (figure 8). Don came highly recommended. His sire had sold for $100,000 and he was truly a champion animal.

On the pretext of further training, a man by the name of Royce misappropriated Don. After several months, he was finally arrested and brought to trial for taking the horse and removing it from the state of Utah. During the period Royce had him, the animal had bred many mares, witnesses claiming a ridiculous number. Royce had kept the horse under the influence of drugs and had given false papers with regard to the breeding certificates.

When the horse was returned and Royce was finally brought to justice, Henry Moray felt such sympathy for Royce - because of his large family and the poverty that Royce professed - that he allowed it to be said in court that the horse was worth less than fifty dollars. This made the crime only a misdemeanor and Royce was not sent to prison. All Henry really wanted was to have his horse back. However, in a way his sympathy backfired because he made an enemy of a man named Grant Ivins, who had had mares bred illegally by Don while the horse was in Royce's possession, and who demanded, unsuccessfully, that Henry give him breeding certificates.

During this period in his life Henry was learning the hazards of fame, not just from people who claimed his prize cock or made off with his champion horse. Before 1928, my mother received phone calls telling her that her husband's life was "not worth a plugged nickel" unless he cooperated on Radiant Energy. Even then, violent opposition to his work was becoming prevalent. His home and laboratory were constantly broken into when the family was not at home.

The harassment became so great that Henry was forced to buy a 32.20

Fig. 7

Lord Cecil with gold medal, a prize chicken in front of the coops, ribbons with silver cup

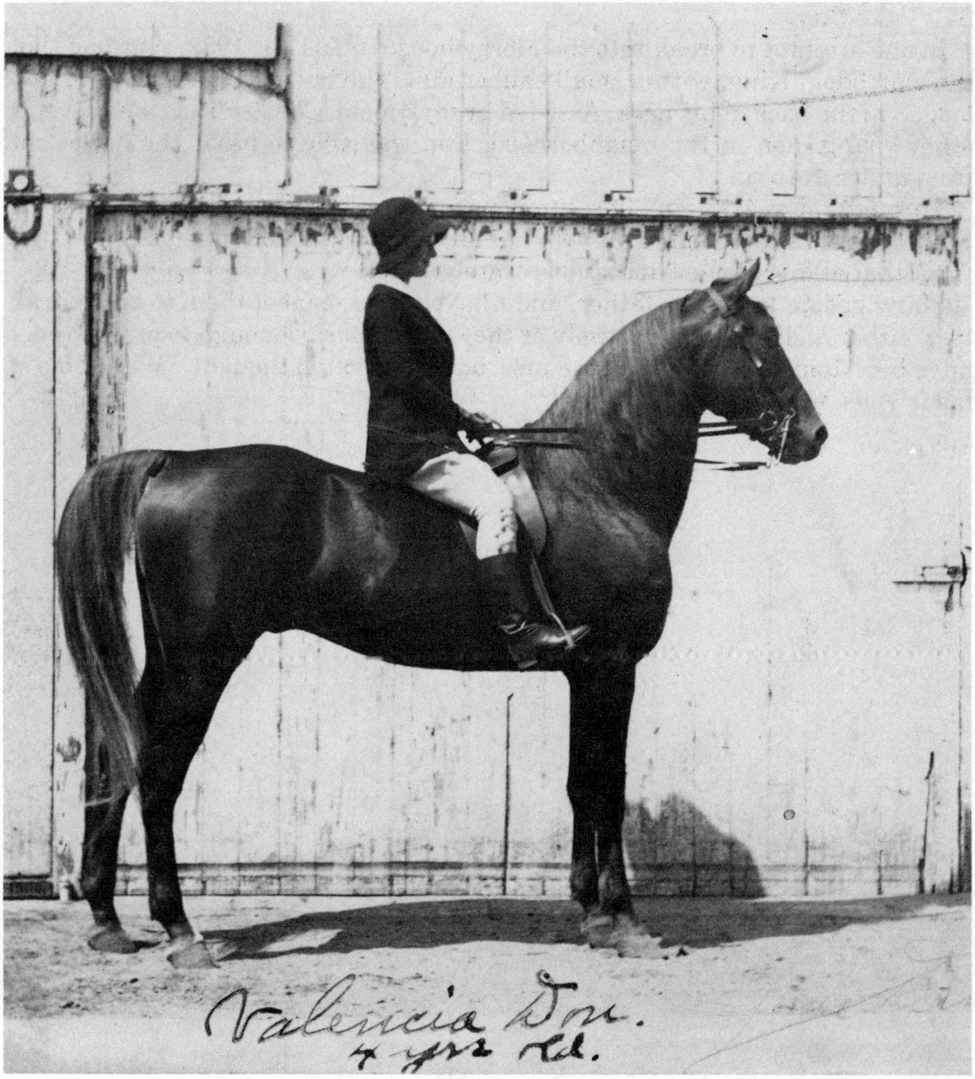

Fig. 8

VALENCIA DON

revolver and a little hand gun, a Colt. 32. For many years he carried the Colt with him wherever he went. Until the federal government entered the picture, Henry always had special bullet-proof glass put in all the windows of his automobiles; a local glass installer had it shipped in especially for him.

More than once the glass saved our lives. I remember one time in 1936 when we were fired upon. My mother was driving the car and I was sitting behind her in the back seat. The bullet went through the car and lodged in the windshield directly in front of her. Just as this happened, a classic black sedan with all the shades down almost forced her off the street and then sped away up 21st South in Salt Lake City.

In one attempt to break into the Moray house in 1931 or 1932, someone shot the watchdog, King, with a small caliber gun. The bullet grazed his head and lodged in the back of his neck. Assisted by my cousin, Chester Todd, and several other young men in the neighborhood, Dad was able to probe the bullet out from under the skin.

Through the years many unscrupulous men tried to deny Henry credit for his life's work or take from him his very livelihood. These actions often hindered the work that all his life he tried so desperately to do.

Many people knew my father, and I have found none of them to be neutral: they either disliked him intensely or they could not say enough to praise him. I have been told that he had no enemies, but I have often thought, "With some of his friends, who needed enemies?"

CHAPTER 2

RADIANT ENERGY IN EARNEST

People, like objects, seem to be governed by the laws of inertia. Ideas or inventions which may change the status quo are viewed with alarm by many scientists, particularly if they cannot understand or explain the new in terms of present knowledge. Through the years, continual experimentation and refinement have gone forward. But always in the long run the work of advancement has fallen back onto one man and his resources.

Many persons have questioned what this thing is that Henry called "Radiant Energy," which has caused so much discussion over the years. In lay terms, it is the direct harnessing of the vast power of the universe to produce electrical energy for man's needs.

Henry Moray demonstrated that energy was available by its actions on a resistive load, such as a flat-iron or a space heater, and by lighting lights. A resistive load is directly proportional to the amount of energy delivered to it. In heating a heater or lighting a light, the number of watts produced can be calculated as being equal to the number of watts consumed. This energy is fed into a load to give either heat/light, or power. A motor can be operated but must be designed for high frequency. The Radiant Energy device operation is shown in figure 54, in which he used an antenna and a ground connected to his solid state R.E. device.

As an innovative researcher, Henry Moray had a dream, a vision, and from his boyhood he pursued this radiant energy. He did not really care where the energy came from. He only wished to demonstrate its existence to the scientific world. He was able to show that none of the energy came from within his device. Internally the device was electrically dead when it had not been connected and tuned to the antenna. When his device was set up, he could connect it to an antenna and a ground, and by priming it first and then tuning it as he primed it, the device would draw electrical energy. This high frequency electrical energy produced up to 250,000 volts and it lighted a brighter light than witnesses had ever before seen. Heavy loads could be connected to the device without dimming the lights that were already connected to it. This device worked many miles from any known source of electrical energy, such as transmission lines or radio. The device produced up to 50,000 watts of power and worked for long periods of time.

Because he was an intuitive researcher, he had built solid state devices using second and third order magnitudes of reasoning, skipping from building a simple crystal set to searching for a detector that would bring in this energy. He assumed at first that this energy was electromagnetic in origin; however, he never claimed that it was electromagnetic in nature. He assumed at first that this energy came from the earth. But later he believed it was from the universe.

Finally he began to believe that it was present throughout all space, intermolecular space as well as terrestrial and celestial space. He did not necessarily understand how his detectors operated, only that if he very carefully built the device according to his calculations, it worked. He was able to demonstrate the existence of an energy that today, though it has not been identified or proven, has been theorized by many researchers.

Henry was not a physicist; he was by training and experience an electrical engineer. The disciplines of the rest of the scientific world meant little to him, which gave rise to some of the difficulty that developed in communication between him and the scientists of his day.

Many attempts have been made to harness the forces of nature to directly produce usable electrical power, and many have failed. Dr. Moray attempted to do it and succeeded, producing at first only a spark, which he developed to a few watts, and finally thousands of watts of power (figure 9).

As early as 1904, Nicola Tesla, experimenting with A. C. currents of high potential and high frequency, said, "Ere many generations pass, our machinery will be driven by a power obtainable at any point in the universe . . . Throughout space there is energy. Is this energy static or kinetic? If static, our hopes are in vain; if kinetic - and this we know it is for certain - then it is a mere question of time when men will succeed in attaching their machinery to the very wheelwork of nature."

Radiant Energy was first presented years before the world was ready to accept it in principle or understand its magnitude. History has again shown the unwillingness of men to accept and adapt to drastic change or new ideas. Henry Moray realized the magnitude of his discovery and felt the burden of responsibility for what could well be the future of mankind. His sincere desire was to place his knowledge where it could do the most good for all, instead of in the hands of a few who were selfish for power and wealth.

As one example of his selflessness, on July 24, 1925, while conferring with Senator Reed Smoot (at the Senator's invitation) in his Salt Lake City office in the Hotel Utah, Henry Moray offered his Radiant Energy discovery to the United States Government *gratis*. Although it now sounds unbelieveable in this day of the serious energy crunch, the senator thanked Moray but stated that the U.S. Government would decline such an offer on the grounds that the government was not competing with public utilities.

Because of conflicting interests, some people refused to become involved. Others deliberately interfered with the development of Radiant Energy. The many offers Moray received for Radiant Energy had to be carefully considered for intent and chances for successful development.

In the early years of its development, the Radiant Energy device was minutely examined during many tests and demonstrations. Some people came to view this device with no other purpose in mind than to prove it a fraud. But no one was ever able to find any evidence of fraud, and all admitted that it was beyond their comprehension.

Through these "demonstrations" Dr. Moray increased his knowledge of the device, and in performing experiments he widened his discoveries, thus im-

Fig. 9

Early RE device at base of antenna. The coils are superficial and used by Henry only to protect his discovery.

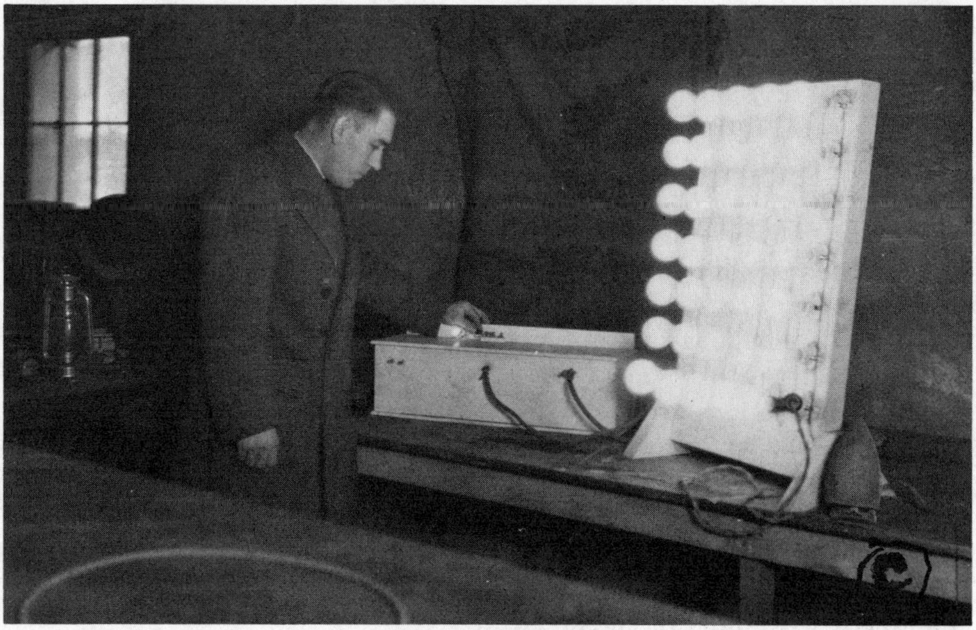

Fig. 10

Dr. T. Henry Moray

proving the device and making it more efficient. On several occasions during these experiments, Henry failed to keep the device working. R. L. Judd's letter, which refers to the experiments for Dr. Harvey Fletcher, states that the device even burned out the detector and Henry had to discontinue his experiments for a time.

The following is a booklet printed about 1932, entitled "Brief History of Moray Radiant Device."

I STARTED my experiments with the taking of electricity from the ground as I termed it, during the summer of 1909. By the fall of 1910 I had sufficient power to operate small electrical devices and made a demonstration of my ideas to two friends, Lawrence and Parnell Hinckley. These demonstrations consisted of operating a miniature arc light. In the early stages of my experimenting I had the idea of using what I termed static based upon the experiments of Benjamin Franklin in his kite but as advancements were made it soon became evident that the energy that I sought was not static and that the static of the universe would be of no assistance to me in obtaining the power that I was seeking.

In spite of the accepted ideas of scientists of the day which all seemed to point to the impossibility of what I was trying to do there was never a time when in my mind, anything but ultimate success was assured. Even when only enough energy was obtained to make a slight click in a telephone receiver I did not doubt the soundness of the ideas upon which I was working. It was during the Christmas holidays of 1911 that I began to fully realize the fact that the energy I was working with was not of a static nature but of a oscillating nature and that the energy was not coming out of the earth but that it rather was coming to the earth from some outside source. That these electrical oscillations in the form of waves were not simple oscillations but surgings like the waves of the sea coming to the earth continuously, more in the day time than at night but always coming in vibrations from the reservoir of colosal energy out there in space. About this time I was able to obtain enough power to light the old type 16 candle power carbon lamp to about one-half capacity and as far as any outward advancement I did not seem to make any until the spring of 1925. There were periods during the years of 1916 to 1921 that I was unable to spend any time upon experimenting owing to the fact of contracts I was under in employment.

In June, 1925, I made certain advancements in my device that made it possible to obtain sufficient current to light a standard type C 100 watt G. E. lamp and made a demonstration of my device to a gentleman by the name of R. L. Shoup and his wife. Some days later to a man by the name of Massey and later to Mr. Massey and a Mr. Brown, who was at that time connected with the Roberts Hotel, of Salt Lake City, Utah. On August 6, 1925, R. L. Judd came down to my home and I let him see my whole device in operation. The largest instrument being about six inches high and circular in shape and about eight inches in diameter. He seemed very much impressed. In October of 1925, Mr. Judd again called at my home upon his return from a trip to New York. We went out on the roof of the chicken coop carrying the device on a small drafting board, erected an antenna upon the roof of the coop, the antenna being about 100 feet from the house. We pulled the main line switches in the house before going out upon the roof. Mr. Judd had me move the drafting board around from place to place and I also had him examine the inside of the coop for hidden equipment. I then put the machine together in his presence and the device was then started. Mr. Judd timed me to see how long it would take to bring in the light. I was able to light the 100 watt type C G. E. lamp to full capacity and also heat to the sizzling point an old styled Hotpoint electric flat iron which consumed 655 watts. Mr. Judd asked me to take off the antenna. The light went out. It was connected again and the light appeared. The same result when the ground was disconnected and reconnected. We drove a new ground at a spot selected by Mr. Judd, made a connection over to the new ground and the light burned dim but came brighter and brighter as the new ground was driven deeper and deeper.

Mr. Judd stayed about two and one-half hours all of which time we had the machine in continuous operation. He wanted to know how long the device would continue to operate. I told him that if he wished to stay and watch all night I would try to make him as comfortable as I could, if he cared to turn night watchman. He said his one great desire would be to have Dr. Harvey Fletcher, of the Western Electric and Bell Laboratories see my device.

After Mr. Judd left I moved the device into the house and kept it in operation all that night and the next day. In fact continuously for three days and three nights and the energy received was as strong at the end of the three days as in the beginning.

In the morning of October 22, 1925, I called upon the General Electric in the Continental Bank Building, at Salt Lake City, Utah, and was introduced to a gentleman by the name of Ship. I stated what I was doing in my experimental work and after some conversation it was arranged that I would show the device to him and to a gentleman by the name of Mr. Dee, that afternoon.

I went home and put the device together. Mr. Ship came to my home with Mr. Dee and I put the machine back of a curtain as I had done in previous demonstrations. After the demonstration I put certain pieces of apparatus in my pocket and then let them examine the rest of the machine as much as they desired. Apparently they could find no fault with what they saw.

The same kind of a demonstration was given to one or two representatives of the Deseret News in August of 1925. They likewise could find no fault with what they saw.

On October 24, 1925, Mr. Judd brought Messrs. Heber J. Grant and A W. Ivins down and I put the machine together before them on the coop, getting off the wet ground as demonstrating on the ground is very dangerous. After going through the same kind of a demonstration as I had given to Mr. Judd the last time of the light and the flat iron, Mr Judd asked me to disconnect the ground. The light went out. It did the same when the antenna wire was disconnected, coming back when the connection was again made. We then drove a new ground as described above with the same results.

Whenever the ground or antenna is left disconnected too long the device becomes electrically dead and must be re-tuned in order to obtain the energy.

On October 28, 1925, I again called on the General Electric and invited Dr. H. T. Plumb, Mr. Ship and Mr. Wheatlake, of that company, to see a demonstration. At the appointed time Messrs. Ship and Wheatlake came to my home and explained that Dr. Plumb had at the last minute found out he could not come to see the demonstration. These demonstrations have all been given at my home, 2484 South Fifth East, Salt Lake City. This time I brought out a small round table on which I carried my device and moved it around to show there were no hidden wires. I then covered the machine to hide the hookup. The machine was started and I lighted a 100 watt lamp and heated the 665 watt flat iron. Removed the ground and antenna wires as described in the other demonstrations with the same result. Mr. Ship and Mr. Wheatlake felt the current after it had been reduced in my transformers and said that it was not direct current and it felt like the current was high frequency. All at once when Mr. Shipp and Mr. Wheatlake were standing by the machine a surge of current came in and caused the electricity to leap across an air gap that we all decided was at least 10,000 volts. The machine acted rather funny and I wished the surge had not taken place as I was afraid that something inside the device had been injured by the surge and that the machine might break down under such heavy voltage on the low voltage side.

By November of 1925, I had given so many demonstrations that I began to lose count of them.

One day Mr. Judd brought Dr. Carl Eyring from the Brigham Young University. I understand Dr. Fletcher had told Mr. Judd that if Dr. Eyring had a chance to see the device that it would be as good as if he, Dr. Fletcher, were to see it himself. Dr. Eyring found no fault with the demonstration and the worst that he could say about it was that it might be induction, but that if I would take the device out in the mountains away from all power lines, a distance of three or four miles, and it would then operate he would then acknowledge that it could not be induction and that his theory was wrong and mine right.

To satisfy the induction theory such a test was made on December 21, 1925, the original account of which has been acknowledged and signed by Mr. Judd, and is as follows:

Dec. 21, 1925:

Today Atty. Judd, O. W. Adams and Atty. Nebeker called at my home in Atty. Nebeker's automobile. When my "Radiant Energy" device was packed in the auto we drove away. The three above mentioned gentlemen then began to discuss where we should go to make the demonstration. I said I did not want to have any say in where we went as I wanted the demonstration made at a place selected by them not me or by my having anything to say about the location. At last they decided to go up Emigration Canyon, as there are no power lines in that canyon. After driving about four miles up the canyon, they selected a place but changed their minds and selected another place a few hundred feet further up the canyon.

Atty. Judd stayed in the car because of an injured foot while Atty. Nebeker and Mr. Adams put up the antenna and ground. I then took the device out of the car and connected it up with the antenna and the ground. The switch on the device was closed and *as in former demonstrations time and time again no light appeared. I then tuned it as in all the former demonstrations that Mr. Judd had seen and then when the switch on the device was closed the light came on. The antenna wire was momentarily disconnected, the light went out but came back when the antenna was again connected to the device. The same happened when the ground wire was disconnected and again connected in the same way as described above. (All this with Atty. Judd hopping around on one foot, he having gotten out of the car when I began tuning the device.) All three gentlemen were very well satisfied and pleased with what they saw. It was dusk when we left the canyon.* * * *

I have read the foregoing and the same is correct so far as I remember.

Sept. 7th, 1929 *(Signed)* ROBERT L. JUDD.

On December 23, 1925, I burned out the machine.

In February, 1926, Attorney R. L. Judd came down with Dr. Eyring, and an account of what took place that day has been acknowledged and signed by Mr. Judd, and is as follows:

February, 1926:

Since the demonstration was made up Emigration Canyon, the device burned up. Atty. R. L. Judd came down with Dr. Eyring at 9 a. m. and we spent all morning going over the theory and hook-up and examining the burned device and the Doctor asked questions continually. He also wrote some notes trying to form a basis for applying for a patent. He made several drawings and sketches in going over the details of the theory on which the device is based. Mr. Judd and Dr. Eyring went uptown at 12 o'clock returning at about 1:30 p. m. After spending all that afternoon with me Dr. Eyring congratulated me on what he termed "my wonderful work" and said what he had heard and seen was mechanically, electrically and scientifically sound and correct. Atty. Judd and the Doctor left in the late afternoon. The next day I called on Atty. Judd at his office in the Kearns Bldg., and he was very much pleased at the stand Dr. Eyring had taken. Atty. Bagley (member of the firm) seemed pleased when Atty. Judd told him what Dr. Eyring had said after having spent the day at my home. My wife was also very much pleased at what she heard Dr. Eyring say.

I have read the foregoing and the same is correct so far as I remember.

Sept. 7th, 1929 *(Signed)* ROBERT L. JUDD.

On May 25, 1926, I had a new and larger machine built exactly like the smaller one only larger and demonstrated in the usual way to Attorney

Judd. We put the device together, he and I. (Judd saw every detail, every part of the machine even to the detector and while he is no electrical man and could not make such a machine he would know batteries if he saw them or be able to see there are no hidden wires which the other demonstrations have proven anyway, especially the one up the mountains.) Mr. Judd and I put up the antenna together and ran the ground. This machine brings in the light quicker than the old one.

May 27, 1926, I again demonstrated the device to Mr. Judd, Mr. Nebeker, Mr. Adams, Mr. E. G. Jensen and Mr. J. W. Knight, going through the same tests of antenna, ground and switch. I have made every test that any man can think of or suggest. Described the theory and made drawings in detail to their best Doctors of Science and not for one word or theory have they been able to find any fault.

In June I demonstrated the device to D. C. Green, C. W. Nibley, R. L. Judd, one of Mr. Judd's brothers, I think his name was Jim Judd, Dr. David Ostler, E. G. Jensen, J. Wm. Knight, Mr. Adams, Mr. Nebeker, Mr. G. E. Ellison, David O. McKay, Dr. Harris, president of the B. Y. U. and others.

In October we went out into the Strawberry Country which is far away from everything in the shape of houses, power wires or telephone wires. This was on October 29, 1926. With me was Mr. R. L. Judd, Mr. E. G. Jensen and Mr. J. Wm. Knight, and I gave them a very splendid demonstration, and on November 18, 1926, I received a copy of the account of this demonstration, written by Mr. E. G. Jensen, copy of which follows:

Salt Lake City, Utah.
November 17, 1926.

Mr. A. C. Cooley,
Salt Lake City, Utah.

Dear Mr. Cooley:

This letter is being written you for your information and in order to make a record of the electrical demonstration made October 29th, 1926, by Inventor T. H. Moray for J. Wm. Knight, R. L. Judd and myself.

As arranged on October 27th, 1926, I met Messrs. Moray and Judd at Mr. Moray's residence, 2484 South 5th East Street, Salt Lake City, Utah, about 8:10 a. m. October 29th, 1926. We loaded the electrical equipment into my car and left Mr. Judd's car in the Moray lot, as there were only three of us to make the trip from Salt

Lake. We had planned taking A. W. Ivins, C. W. Nibley, D. O. McKay and James E. Ellison, but none of these gentlemen could go.

We arrived at Orem, on Provo Bench, about 9:00 a. m. and just a few minutes before Mr. J. Wm. Knight arrived from Provo. After transferring the electrical equipment, lunch and rubber chair mat into Mr. Knight's car we started for a location to demonstrate, roughtly in the vicinity of Strawberry Lake, the object being to get far enough away from all power lines to demonstrate that the current used was not induction from power lines.

I remember that the speedometer registered 19 miles at Charleston and 26 miles as we left the last electric line near the mouth of Daniels Canyon; also that the speedometer registered 52 miles where we stopped to demonstrate, making the distance 26 miles from the nearest power line.

** * *As we neared the summit the sun showed through the clouds at times and as we proceeded down Strawberry Valley the weather conditions improved so that Mr. Moray readily consented to make a demonstration at any place we might select. His only concern seemed to be that he did not like to demonstrate in a storm on account of the danger involved.*

Mr. Moray requested that we select a place near a stream of water so that the ground pipe could be sunk in its bed and be more effective. We stopped at a place about 10 miles southeast of the Daniels-Strawberry summit and about 200 yards west of the main road to Duchesne, this location being almost due east from what Mr. Knight called Haystack Mountain and perhaps ¾ of a mile east of the Strawberry Lake on a little stream which made a zig-zag course through a gently sloping, grassy flat to the lake. * * **

The antenna wire was put up without any aid or instructions whatever from Mr. Moray, except that he suggested that the wire be stretched tighter to prevent so much sag at the center. This was done and the wire then appeared to clear the ground by about 7 or 8 feet at its lowest point.

The ground pipe was of ½-inch water pipe consisting of two sections. The lower section was pointed at the end to make its driving into the creek bed easy. It was about 6 feet long and after being driven down about 5 feet the second section, which was about 4 feet long, was screwed on with a Stillson wrench and the pipe further driven down until it struck a hard object and began bending near the top . I judge about 7 feet of pipe was in the ground.

The antenna wire was insulated from the poles with two glass insulators about 6 inches long and having holes in both ends. A piece of wire about 2 feet long connected each insulator with the pole. The lead-in wire was fastened to the antenna wire at a point about 10 or 15 feet from the east pole. I helped Mr. Moray solder the connection where the lead-in wire fastened on to the antenna wire and also helped him solder the ground wire to the pipe. I stepped the distance between the two antenna poles and estimated it to be 87 feet as I took 29 steps intended to be 3 feet each.

Mr. Moray's equipment, aside from the antenna and ground wires, consisted of a brown box about the size of a butter box, another unpainted box slightly smaller, a fibre board box about 6" x 4" x 4", which Mr. Moray referred to as containing the tubes, and a metal baseboard about 14" x 4" x 1" containing what appeared to me to be a magnet at one end, a switch near the middle and a receptacle for an electric light globe at the other end. There were also several posts for connecting wires on the baseboard.

Mr. Moray took this electrical equipment out of the automobile and placed it on the running board of the car. Two dry boards, which Mr. Knight brought, were laid on the ground and a rubber mat used under my office chair was placed on the boards for Mr. Moray to stand on as a precaution against electric shocks. The running board was hardly large enough for the equipment so we took the seat cushion out of the front seat and placed it on the mat and Mr. Moray transferred the equipment to the seat cushion and connected it up there.

Very light snowflakes fell occasionally and a tarpaulin was hung over the top of the auto doors, when opened, to protect the equipment from getting wet. When all of the wire connections were made and everything in readiness Mr. Moray began tuning in. It was just 1:05 p. m. by my watch. Before tuning in he placed the key on the post he said it would be in contact with while the light burns but no light appeared. The tuning in consisted of stroking the end of a magnet across two metal projections protruding from which I referred to above as being a magnet. After tuning in for slightly more than 10 minutes the key was put on the operating post and the light appeared immediately. It was slightly after 1:15 p. m. by my watch. Mr. Moray put the key on the operating post two or three times before during the tuning in operation but no light appeared. We allowed the light to burn for 15 minutes, or until 1:30 p. m. The brilliancy of the light, a 100-watt globe furnished by Mr. Knight, was about 75% as bright as a 100-watt globe connected to an ordinary house socket in my opinion. It was an even light, without fluctuations of any kind.

While the light was burning Mr. Moray disconnected the Antenna lead-in wire from the apparatus and the light went out. He connected it again and the light appeared. He also disconnected the ground wire and the light went out. He then connected it and the light appeared again.

Mr. Moray seemed confident that he would be able to obtain a light and went about his work in an unhesitating, skillful manner. He said he knew he could do the same thing in the middle of the

*Sahara Desert or in the deepest mine. When the demonstration was over we congratulated Mr. Moray and I felt confident that he had a real invention and that no hoax was being perpetrated. * * **

Yours very truly,

(Signed) E. G. JENSEN.

In the spring of 1927, a gentleman from Massachusetts, Mr. H. F. Haffeneffer, Jr., one of his sons, and one of Mr. Haffeneffer's engineers, together with Mr. Judd, Mr. Jensen, Mr. Knight and Dr. David Ostler witnessed a demonstration of my device. When all was over and the usual tests of taking off the antenna and ground and throwing the switches and the light had burned and the flat iron heated to their satisfaction for a long time, I asked them if they had seen enough. When they said yes, I opened all of the device and let them see everything except one small part that I was able to place my hand over and hide in my fist. This part I cut off and put it in my vest pocket. Everything else I let them examine to their heart's content. "If that part is able to make such power itself its some device and worth selling. Such a battery would be worth while," were some of the remarks passed.

On September 20, 1928, I gave Dr. Fletcher every opportunity of studying in detail the drawings, demonstrating the circuits involved and the theory upon which the mechanism is based and on Tuesday, September 25, I gave Dr. Fletcher a complete demonstration, report of which is made in a letter to Mr. H. F. Haffeneffer, Jr., by Mr. R. L. Judd, and is as follows:

September 27, 1928.

Dear Mr. Haffeneffer:

** * * We have though just finished a most interesting demonstration that I am sure you will be pleased to hear about.*

** * * on Thursday a week ago Mr. Moray and Dr. Fletcher spent the afternoon in a detailed study of the drawings demonstrating the circuits involved and the theory upon which the mechanism is based. This was as requested by Dr. Fletcher. On Tuesday last Dr. Fletcher, myself and Mr. Jensen, another interested party, went down to Moray's home. Dr. Fletcher was afforded the opportunity of studying the antenna, the ground and their respective connection with the machine. Moray then demonstrated that there was no life in any part of the machine, and explained to the Doctor what he was about to do. He then began his tuning in as you have seen him, and in six minutes from the time of his commencement had sufficient power to light three one hundred watt lamps. Dr. Fletcher in observ-*

ing the lamps stated that they were burning far above normal and that if they were to continue at that degree of radiancy they would soon burn out. While observing the lights various tests were made such as disconnecting the antenna and the ground. The lights were then screwed out quickly and a five hundred seventy-five watt flat iron attached. This was heated to the "sizzling" point in about five minutes. After making such observations as were necessary to this test, the wire was disconnected and the lights again put on. After the lights had been burning for some little time, for some reason, then unknown, they flickered a few minutes and then went out. The lights on the regular circuit in the house still going on. The globes were tested and found to be all right. A later examination of the parts of the machine showed that one part of the detector, that piece of mechanism which is not yet "fool" proof, had slipped down out of place and in affect had disconnected that particular circuit. This happening pleased Dr. Fletcher and was considered hy him a valuable part of the experiment. After the lights had gone out the mechanism was dismantled part by part and the Doctor given a chance to see that the mechanism was one intrical whole with no other connection than with the antenna and the ground.

His conclusion was without further time or equipment with which to study the circuits, and the extent of them going through different parts of the machine, that the energy secured by the machine was what Dr. Moray contended for it or that Moray secured from his antenna and ground a force which set up in his tubes, the makeup of which were explained to the Doctor, some form of battery action which resulted in the electric current generated. He stated that if it were the latter however, the phenomenon was quite as remarkable as contended for by Moray, because in accomplishing what they did the tubes were far more powerful than anything known to science today.

His observation as to the possible weakness of the thing if it was some battery reaction in the tubes which was commenced and kept alive by some form of energy from the atmosphere, was that the tubes would soon burn out and that it would be necessary to do further work with the tubes to give them long life. I told him that the tubes built two years ago last June had done service up to the last month when Moray, through their leaking on account of what he says is his crude way of making them, had had to make new ones. His answer to that was that if it is battery action the tubes might be used for sometime then left idle that they would rebuild themselves and thus go on serving as the tubes mentioned had done. He told us that an experiment that would be very helpful in determining what the energy was or the efficiency of the tubes would be to run the machine just as long as it would go, then determine what part quit first and if it were the tubes make observation as to what happened. I didn't know just what Moray's response to that suggestion would be, but yesterday he dropped in to say that he was very anxious to make the test as soon as possible, and asked that I assist him in figuring out the proper personnel of a committee, seal the machine up in some way after it was lighted, and then make frequent observations as to its continued operation for the full period of time it would run. I expect to work out such arrangements as soon as possible.

From this demonstration it is established, I should say absolutely, that what we have is something entirely new and very wonderful and that there is no possibility for doubt on any phase of this proposition.

With kindest regards, I am

Yours respectfully,

(Signed) Robert L. Judd.

The following is another report on a demonstration:

Salt Lake City, Utah,

September 26, 1928.

Mr. R. L. Anderberg,
Los Angeles, California.

Dear Reed:

This letter is being written for the double purpose of informing you of an electrical demonstration given by Dr. T. H. Moray to Dr. Harvey Fletcher, R. L. Judd and me and also for the purpose of making a record of same. The demonstration took place Tuesday, September 25, 1928, in Moray's laboratory in the basement of his home and consisted of his lighting three ordinary electric light globes of 100-watt capacity each. He also heated a 575-watt flatiron. The lights were somewhat brighter than when placed in an ordinary lighting socket supplied by local utilities energy. It took about six to seven minutes to tune in and get the light, and after demonstrating for possible eight or ten minutes as above referred to the lights went out. This was due to the detector getting out of adjustment according to Dr. Moray. After the demonstration Mr. Moray took the covers off the boxes containing the equipment and disconnected the wires connecting the several pieces of equipment and allowed us to handle all of the equipment except the detector, which he let us see as he held it in his own hand but did not let us examine it closely or handle it.

After the demonstration, which was preceded by a conference between Dr. Fletcher and Dr. Moray sometime before in which the circuits and hookups were explained, Dr. Fletcher remarked that it was a wonderful demonstration and just about what he expected to see after having had it explained to him in the preliminary conversation.

The Doctor said there was nothing much to say about the apparatus, but that if it were turned over to him he would soon find out much about it that he did not then know. He said the energy came from one of two sources: either an external source, as claimed by the inventor, or possibly through harmonizing or synchronizing the various units comprising the equipment through his tuning in process in such a way that a chemical action resulted in the tubes sufficient to make the demonstration. In either event he agreed it would mean a wonderful discovery. The former source of supply was the one thought most probable according to Dr. Fletcher.

** * *The test was the same as all others I have seen, including the one out on the Uintah Reservation near Strawberry Reservoir. The same tests were made of disconnecting the antenna and the ground wires and in each case the lights went out. When disconnecting these wires a bright spark of considerable length could be seen, which was referred to as a "brush spark" and Dr. Fletcher stated the power was "high frequency."*

The large antenna was used and the lights were brighter than I have ever seen them at former demonstrations. We were in the laboratory from one and one-half to two hours and the demonstration and conversation between Dr. Moray and Dr. Fletcher was very interesting. * **

Yours very truly,

(Signed) E. G. JENSEN.

The following are reports of endurance tests made:

Salt Lake City, Utah.

October 8, 1928.

Mr. R. L. Anderberg,
Los Angeles, California.

Dear Reed:

On September 26, 1928, I wrote you regarding a demonstration given by Dr. T. H. Moray to Dr. Harvey Fletcher of New York, R. L. Judd and the writer, of the equipment whereby Dr. Moray reduced cosmic energy to electrical energy for commercial purposes.

The letter made mention of a suggestion by Dr. Fletcher that an endurance test to determine how long a light will burn would be valuable information to have. Acting on this suggestion Dr. Moray has made such a test.

The test was started on October 1, 1928, and was conducted in Dr. Moray's laboratory in the basement of his residence, 2484 South 5th East, this city. The equipment, which was the same as I have seen on several previous demonstrations, including the one made for Dr. Fletcher, was enclosed in two wooden boxes, which were in turn placed in a trunk having two holes bored in it to admit connecting ground and antenna wire to the equipment and two additional holes of about one-half and three-quarter inch diameter respectfully for ventillation and observation purposes.

Dr. Moray began tuning in at 7:49 a. m. and switched on the light at 7:59 a. m. Two globes were used, a master globe of 100-watts capacity and a pilot globe of 10 watts capacity; the purpose of the two lights being to insure continuous burning of one at least, even though the other should fail. The trunk was closed and sealed immediately after tuning in and in the presence of Dr. T. H. Moray, Dr. Murray Hayes, Mr. R. L. Judd and the writer. Railroad seals of the foolproof automatic locking type were used in sealing the trunk. They were applied on three different places and an accurate record of their numbers and locations was kept by the writer. The trunk was the ordinary wooden construction reinforced with sheet iron.

It was agreed that the three of us, not including the inventor, should visit the laboratory as frequently as we could conveniently do so to see if the lights were still burning and that the equipment had not been tampered with and to observe the brightness of lights and any other things pertinent to the test.

The following is a record of the inspection made by me:

Date	*Time*	*Lights*
October 1, 1928	*7:59 a. m.*	*Burning O. K.*
October 1, 1928	*6:30 p. m.*	*Burning O. K.*
October 2, 1928	*8:50 a. m.*	*Burning O. K.*
October 2, 1928	*6:35 p. m.*	*Burning O. K.*
October 3, 1928	*8:40 a. m.*	*Burning O. K.*
October 3, 1928	*6:30 p. m.*	*Burning O. K.*
October 4, 1928	*7:50 a. m.*	*Burning O. K.*
October 4, 1928	*10:20 a. m.*	*Burning O. K.*

Time of continuous burning 74 hours, 21 minutes.

Dr. Moray telephoned about 11:00 a. m.. October 4, 1928, that the light was out. He stated that large poplar trees were being topped near his laboratory and that in dropping to the ground the tops shook the ground sufficiently to throw the detector out of adjustment and stop the light. I was standing in front of Mr. Moray's house when the top of one of the large trees fell and know that a tremendous vibration of the ground, which, which is somewhat boggy, took place.

Dr. Moray suggested that the three witnesses arrange a convenient time to meet and the laboratory, unseal the trunk, inspect the apparatus and decide on further procedure.

At 6:30 p. m., October 4, 1928, with Moray, Hayes and Jensen present, seals were inspected and found to be O. K. Seals were then broken, the trunk lid raised, and the cover to the top box unscrewed and taken off and the detector only taken out. Mr. Moray shook the detector gently and we all heard a rattling sound, which Mr. Moray pronounced as the part of the detector jarred out of position when the trees fell. Mr. Moray further stated that he thought he could adjust it quickly and started to do so immediately in the laboratory and in our presence. The detector was pronounced O. K. and ready for installation and further demonstration at 6:53 p. m. We left Dr. Hayes in the laboratory to watch the equipment while Moray and I went upstairs to telephone to Mr. Judd who advised that he would come immediately to watch the tuning in and sealing of the trunk. Judd arrived at 7:35 p. m. and tuning in started immediately. The light was obtained at 7:44 p. m. Judd's two sons were present also. The trunk was sealed again the same as before and an accurate record of seals taken.

A further record of inspections by the writer follows:

Date	*Time*	*Lights*
October 4, 1928	*7:44 p. m.*	*Burning O. K.*
October 5, 1928	*8:20 a. m.*	*Burning O. K.*
October 5, 1928	*7:15 p. m.*	*Burning O. K.*
October 6, 1928	*8:20 a. m.*	*Burning O. K.*
October 6, 1928	*10:00 a. m.*	*Burning O. K.*
October 6, 1928	*7:55 p. m.*	*Burning O. K.*
October 7, 1928	*9:10 a. m.*	*Burning O. K.*
October 7, 1928	*10:55 a. m.*	*Burning O. K.*
October 7, 1928	*5:25 p. m.*	*Burning O. K.*
October 8, 1928	*7:18 a. m.*	*Burning O. K.*

Time of continuous burning 83 hours, 34 minutes.

It was decided late Sunday night that the test should be discontinued Monday morning, October 8th, and accordingly Messrs. Moray, Hayes, Judd and Jensen were at the Moray laboratory at 7:00 a. m. to conclude same. All seals were examined carefully and found to be O. K. The trunk lid was opened and Mr. Moray made a demonstration by heating a Hotpoint flatiron, No. N. P. 37-J, catalog 115, F 51, watts 575, volts 110, made by Edison Electrical Appliance Company, and at the same time as an additional load he lighted a 60-watt lamp.

Dr. Moray demonstrated further by taking the antenna wire off and allowing the light to go out at 7:18 a. m. and after a brief interval we tested it and found it to contain no electrical energy. Mr. Moray then tuned in and got the light again at 7:22 a. m. He then put on five 100 watt lamps, all of which were lighted brightly, then disconnected and lost the light at 7:24 a. m. After an interval of a minute he started tuning in again and got the light in one minutes time and again lighted the five 100-watt lamps. While these were burning, Dr. Moray jarred the workbench on which the apparatus stood by hitting it a moderate blow with a hammer. The lights flickered a time or two and then went out and it was impossible to bring them back by tuning in in the regular manner or otherwise, thus indicating that it was the jar of the falling trees that was solely responsible for the lights going out on October 4. During the period of the test Messrs. Hayes and Judd made inspections almost as frequently as I did. On several occasions Dr. Moray would disconnect the antenna wire momentarily, but not long enough to lose the light. In disconnecting and connecting the antenna wire a flash of electricity could always be seen at the connecting point. On one occasion Dr. Moray turned the regular lights on in the basement, then went upstairs and opened the mainline switch that cuts all house lights off. The lights in the trunk continued to burn while the others were out.

During the entire test the lights burned evenly and brightly without flickering and there was no change in the brilliancy noted from day to day. It was noted that the 100-watt, type "C," globe bought new for the test, had a dark spot in the glass opposite the filament when it was taken off, indicating that it would not last much longer. Also a slight rattle could be heard in the globe when it was shaken lightly near one's ear. The globe, however, was still good as it lighted when attached to a regular lighting circuit.

After these demonstrations the equipment was taken out of the trunk, the wires disconnected and the respective parts, with the exception of the detector, examined and handled by us.

Dr. Moray made one demonstration not mentioned above to the writer while he only was present. It consisted of lighting a 100-watt globe from connections with the antenna wire only. It was noted that while this light was burning the lights inside the trunk burned dimly and then assumed their usual brightness when the other lights was taken off.

The test was a wonderful demonstration. The inventor was frank and open in all things and had us inspect carefully and to our satisfaction that there were no hidden wires or fake connections and I feel positive there were none.

Yours very truly,

(Signed) E. G. JENSEN.

Salt Lake City, Utah,
October 10, 1928.

TO WHOM IT MAY CONCERN:

Below is a report of my observation of the Moray cosmic energy device.

The aerial used is about two hundred feet long and is about eighty feet above the ground; the wire is a copper cable approximately a fourth inch in diameter, and well insulated. The ground is the water pipe in the basement of Dr. Moray's home.

The device was assembled in a trunk through the sides of which were holes for the connections to ground and to the antenna and for observation; the said holes were about one-half inch in diameter. There were two boxes about ten by twenty by four inches, one on top of the other; both were closed and the covers fastened with screws. On the upper box was lying an insulating panel about an inch thick by fifteen inches long and three inches wide; it is of slate or hard rubber or some material of similar appearance. On this were two binding posts that were connectible by means of a small switch; also mounted on this panel is a body about two and one-half inches square, wrapped in friction tape, from which protrude two poles about one-fourth inch in diameter, apparently of soft iron. A double resceptacle for light globes was connected in the circuit, in one of which was a twenty watt globe, and in the other a hundred watt globe.

E. G. Jensen, R. L. Judd, and I were present and examined the trunk to see if there were any connections other than to the antenna and the ground, but found none. The small switch above mentioned was thrown several times, but without results; the connections to ground and antenna were also removed, also without result.

Dr. Moray then took a magnet, which was a very broad, short limbed U, and began to stroke one pole of it on the poles in the taped body; Mr. Jensen placed his fingers on the binding posts several times, and at last received a rather vigorous shock; Mr. Moray then threw the switch and the globes lighted. When the switch was opened the lights went off, and came on again when the switch was closed. Removing either the ground or the antenna connection caused the lights to go out, but they came on again as soon as the connection was re-established. The time of excitation was ten minutes, and the lights came on at 7:59 a. m., October first.

The trunk was then closed and sealed with railroad car seals, and the numbers on the seals were recorded by Mr. Jensen. Each morning and night up to the morning of Oct. 4, I inspected the seals and observed that the lights were burning. About 10:30 on this date the detector was jarred out of adjustment by the falling of a heavy tree next to the house.

The evening of the same day Dr. Moray removed the detector in the presence of Mr. Jensen and me and in twenty minutes had it readjusted and reassembled ready to start. As soon as Mr. Judd arrived the stroking began and the lights came on in about ten minutes. The trunk was again sealed as before.

The device continued in operation until the morning of Oct. 8, and the trunk was opened in the presence of the three witnesses above mentioned after a run of eighty-four hours. The hundred watt lamp was removed and a standard 575-watt Hotpoint electric iron was plugged in in its place; the iron was heated as quickly as though on the usual house circuit. During this test a sixty-watt globe was put in the place of the twenty, so that the total wattage was 635.

The antenna and ground connections were then removed until no current was delivered when the switch was closed, and five 100-watt globes were substituted for the iron, making a total power output of 560 watts. The lamps appeared to be as bright as when on the house circuit. It required four minutes of excitation to get it in operation again.

After again being disconnected until it ceased to operate it required but one minute of excitation to bring in the current.

While the test was in progress every test that could be thought of was applied to make sure that there were no hidden connections to the house circuit or to a battery; the house lights were turned on and then all main switches pulled, which turned off the house lights but did not in the least affect those in the test. After the run had terminated the trunk and table were examined for wires, but none were found except those of the antenna and to the ground.

As a further proof that the conversion of the energy was due to the mechanism in the box, Dr. Moray hit the table on which the trunk was standing, a moderate blow with a hammer whereupon the light flickered and went off, due to the detector being shaken out of adjustment.

The boxes in which the mechanism had been housed during the test were opened and the contents examined; there were condensers, the detector, a transformer, and two tubes in them but nothing else. Nothing that in the least resembled a battery.

It is to be noted that after a total run of 158 hours the device supplied 635 watts; inasmuch as a horsepower is but 746 watts this equals 0.878 of a horsepower or slightly more than ⅞ horsepower. This alone is sufficient to dispose of any suggestion of a battery.

In witness to the above I hereunto sign my name.

STATE OF UTAH } *ss.*
COUNTY OF SALT LAKE }

Dr. Murray O. Hayes, being first duly sworn, deposes and says:

That he has read the foregoing statement and acknowledges that he wrote and signed the same as above set out.

(SEAL) *(Signed)* MURRAY O. HAYES...

Subscribed and sworn to before me this 29th day of October 1928.

(Signed) Manola Jorgensen.

Notary Public residing at Salt Lake City, Utah.

——————

October 10, 1928.

Dr. Harvey Fletcher.
Bell Telephone Laboratories,
463 West Street,
New York.

Dear Dr. Fletcher:

I received your letter of recent date and was pleased to learn that you had arrived home safely after having an enjoyable trip across the country. I thank you for your observation relative to Moray's machine.

Moray seems to have been favorably impressed with his visit with you for the day after you were there he came up to my house, where I was working on a brief in order to not be disturbed as I am at the office, and said that he would like to follow out your suggestion of an endurance test. In order to carry out the suggestion at once we arranged on Saturday with Jensen and Dr. Murray O. Hayes to start the test on Monday morning. Accordingly Monday morning, October 1st, 1928, at 7:30 o'clock Dr. Hayes, Jensen and myself met at Moray's house. We went to the basement where you saw the machine and found that Moray had set up his machine in a heavy tin covered trunk with a small hole in the top and another in one end through which the globes attached to the machine could be observed. The ground hook-up is on the inside of the trunk—the antenna hook-up just outside the trunk. Moray, before starting to tune in, made all the tests as to possible current in the machine. After completing the tests in the presence of the three men above named, he began his tuning-in which lasted about seven or eight minutes before he switched on the lights which consisted of one 100-watt and one 10-watt, this was at 7:59 o'clock Monday morning.

After the lights had burned for a short time Moray extinguished the lights by unhooking his ground connection, but reconnected the ground before his current was lost. He then performed the same test on the antenna connection. When these tests were completed the trunk was closed and locked, the key being delivered to Dr. Hayes who holds the same. The trunk was then sealed in three different places with standard freight car seals which all bear numbers. This trunk cannot possibly be opened without breaking these seals and after they are broken they cannot be reused.

After taking every precaution we could think of we left with the lights burning brightly.

I returned to make my next observation Monday evening at 6:10 o'clock as I was returning from a game of golf at Nibley Park. Moray had no notice of my coming and he stated that I was the first of the three men mentioned above to make an evening visit. I found all the equipment in the exact condition as we had left it in the morning and the lights burning brightly. The weather all day had been clear and pleasant.

Tuesday morning, October 2, 1928, I picked up Jensen and we drove down to Moray's, arriving there about 8:30 o'clock. We found everything in status quo with the lights burning brightly. Moray repeated a test he had shown me the night before of disconnecting the antenna hook-up thereby extinguishing the light and then reconnecting again and bringing on the light. We stayed on this visit about fifteen minutes.

Tuesday evening about 7:00 o'clock I drove down and made a careful observation of everything. I found it intact with the lights burning as usual. Moray shut off the basement lights and then later pulled the switch on the entire house but neither operation effected the lights in the trunk. In the morning it was fair but during the afternoon there had been some south wind so that by evening the sky was entirely clouded over with a little dampness in the air. Both Hayes and Jensen had preceeded me on this visit.

Wednesday morning, October 3rd, 1928, I picked Jensen up as I had the morning before and we arrived at Moray's about 8:40 o'clock. We made the usual observations and found everything intact with the lights burning as usual. Moray at this time disconnected the antenna hook-up several times and each time the light would go out and come on again when the reconnection was made. He at this time also repeated the tests of pulling the basement and house switches which put out the lights on each of those circuits but in no way affected the lights in question. The morning was very threatening although it was not storming.

In the evening I was unable to get down but talked with Moray who informed me that Hayes and Jensen had both made their visits and had found the lights burning brightly. The evening was rather stormy, it having rained during the afternoon. The temperature was rather cool.

Thursday morning, October 4th, 1928, I visited the machine about 8:15 o'clock and found everything in satisfactory condition. Moray was not at home so I made the examination alone. Shortly before noon of the same day Moray called up and said that the light was out; that it occurred apparently at the time his neighbor immediately to the north had felled a large poplar tree which had jarred the ground considerably. He stated further that it was certain the only trouble was in the detector and that if we would come down and unseal he trunk he would take that particular piece out and fix it and tune in again. He also called Dr. Hayes and Jensen who got down to his place about 7:00 p. m. Together they unsealed the trunk, watched him remove the detector only, reset it and then called me. I got down about 7:30 p. m. Moray shortly thereafter tuned in and at 7:44 had the lights buring again. After making several tests with the antenna, the gorund and different lights, the trunk was again sealed and we left.

Friday, October 5th, 1928, I visted the light twice and found it intact.

Saturday, October 6th, 1928, I visited the light about 1:30 p. m. with J. C. Martin of the Insul Electric and Gas Companies of Chicago, and E. A. Angley, their local manager here. We found the light going on as usual. The men present checked on the connections outside the trunk and talked to Moray about theories and developments.

Sunday morning I was down about 9:30 o'clock and found the light burning as usual. The weather during all these last four days was fair and rather warm considering the season of the year.

Monday morning I was at Moray's at 6:50 o'clock. Dr. Hayes and Jensen arrived a little later. Moray stated that on account of his wife's (illness) he wished to end the test, but before doing so wished to make several tests showing the capacity of his machine. Accordingly after checking all seals to find they were intact, we broke them and let Moray into the machine. He put on an extension cord so that he could heat the flat iron (575W) and light a globe at the same time. This was done with the use of a 60-watt globe, making a total pull on the machine of 635 watts. He then attached four 100-watt globes which were lighted brilliantly. The antenna was then disconnected and the machine was allowed to die. Moray then tuned it in again in about four minutes. It was allowed to die again and then tuned in again in about one and one-half minutes. After having been there for a considerable time and gone through with these various experiments, Moray to demonstrate the delicacy of the detector hit the bench the trunk was sitting on with a hammer, the light flickered about twice and then went out. The machine was then taken apart and each part looked at as at the time of our demonstration with you.

I am enclosing you a copy of Jensen's report of the test.

With kindest personal regards, I am

Sincerely yours,

(Signed) ROBERT L. JUDD.

The following are a few comments received concerning this device:

June 5th, 1929.

TO ALL WHOM IT MAY CONCERN:

It is now more than two years since I first became acquainted with Dr. T. H. Moray and the work he is carrying on, and in that time he has demonstrated inventive ability of an exceptional order.

Perhaps the most wonderful of his inventions is a device whereby he is able to draw electric power from an antenna. This energy is not derived by induction from power lines, as has been suggested by some, nor is it derived from radio stations, as has been demonstrated by taking the apparatus more than 26 miles from the nearest power line and over a hundred miles from the nearest radio station and showing that it operates just as well as anywhere else. This device was subjected to an endurance test in which it was operated continuously for a week, and at the end of that time a 100 watt lamp was lighted simultaneously with the heating of a 575 watt standard Hotpoint flat iron, making a total of 675 watts; it is very evident that no batteries could sustain such a drain as this.

He has also invented a very sensitive sound detector whereby it is possible to hear conversations carried on in an ordinary tone of voice at a distance of several blocks.

He has also worked out numerous radio hook-ups which eliminate many of the parts now considered necessary for good reception, yet there is no apparent diminution in quality or volume; in fact, there is a notable elimination of interference from static when some of these are used.

He has devised a means by which he is able to measure with some degree of accuracy the energy evolved during mental activity; that is, he gets definite, variable deflections of the needle of a sensitive galvanometer which appeared to be related to the vigor of mental activity.

There are a great many other equally remarkable things that he has done, such as reducing old rubber from truck tires to the state of a viscous fluid which is readily vulcanizable without the addition of smoke sheet as is necessary with other processes; also a high frequency theraputic device, and numerous other devices which show great ingenuity.

(Signed) MURRAY O. HAYES, *Ph. D.*

— — — — — —

October 7, 1929.

Dear Dr. Kinsley:

Recalling the pleasant afternoon spent in the laboratory of the New York University with Mr. W. H. Boehm and yourself, Wednesday, September 25, 1929.

You asked some questions at that time that I could not answer on account of my lack of knowledge, both of the science and the detail construction of the machine of Dr. Moray's that we were talking about. There was one thing in particular that you asked about and that I know I did not make clear in my answer, and that was in regard to the set-up of the six 100 watt lamps. You asked if they were set up in multiple or series, and this question was especially brought up as we were referring to the very small wire, which is a No. 50, which I stated was used in connection with the operation of said lamps. Since my return to Salt Lake, I have gone over this matter and find that the six lamps are set up in parallel and the small wire is used as the current enters the tube prior to and connecting with the step-down transformer, this taking the terrific high voltage to the transformer. In other words, I am trying to say to you that this is a terrific high voltage, and we are using the word "terrific" because it has been demonstrated that this is an exceedingly high voltage because it will through an arc flash at least six inches and in excess. * * *

Another thing that you asked me about was the measuring of the voltage or ampherage, and I know I did not answer to your satisfaction. However, I must now state that "the frequency is so high that I have no instrument in my laboratory that is able to measure the ampherage or the voltage at this high frequency." * * *

We have, at Salt Lake City, a Dr. Murray Hays, Ph. D., who has in addition to being instructor in the University from which Dr. Harvey Fletcher was taken to Bell Laboratories, has qualified as a Patent Attorney, and has, on this account, done a considerable amount of research work and understands and realizes the importance of detail investigation for the presenting of an application for a patent, etc. Dr. Murray Hayes has examined most carefully and knows of the construction of Dr. Morey's machine, knows the theory in detail, has practically prepared the papers for patent applications, and which are being held as per my conversation with you because it has been the desire to perfect one small item of the same, and Dr. Murray Hayes tells me positively that there is a new basic, there is a true basic and a presentation that has never been made by any other presentation that he has seen or heard of up to this time. In fact, Dr. Murray Hayes has analyzed with me the presentation made by the Italian to take power from the atmosphere or Thermo Electric effect. He has also gone over the Nikola Tesla demonstrations, and I feel that he knows whereof he speaks, and my reason for putting this part of the letter before you is an inquiry as to whether a presentation in detail by Dr. Murray Hayes would not be as valuable as presenting the whole detail again to such a person as Dr. Parnot.

There is one question that you asked, or at least I have in my mind the question having been considered in our conversation at your laboratory and that is, could the machine be increased in size and the results be increased in volume so that the requirements of *such an undertaking would be fully or more heavily realized. I think the best answer to this question is that the first machine gave a result only by a noise to indicate that there was a power transmission being brought about; the second machine resulted in the lighting of a very small lamp; the third machine was used to make a demonstration of lamps equal to 100 watts; and the present machine, which has been completed recently, will carry six 100 watt lamps and to the point as described by the comments above in regard to Dr. Fletcher. Therefore, I have every reason to believe that with the proper facilities for construction that we could positively make these machines to any size that might be desired. * * **

Yours truly,

(Signed) W. H. LOVESY.

September 2, 1930.

Research Corporation
New York City, New York

Gentlemen:

*Some little time has elapsed since my call on Dr. Poillon and interested interview with your Mr. H. B. Keppel, Jr., in your offices at New York City. * * **

Mr. Moray moves very cautiously and consequently sometimes very slowly. He has had many inquiries and propositions made of and to him—and some of them have been of nature requiring he weigh carefully and study deeply the intent of the proposing parties.

*Many of these presentations and offers come through the friends of the friend of the friend, and call for investigations, sometimes call for "Stops"—to "Look and Listen." Many times we see, or we should say we feel we can see the "fine Italian hand" of a gigantic concern trying to "tie up". and not develop this marvelous invention. * * **

The reason for deciding that no more or promiscuous demonstrations are to be made should be obvious—and, however, we may add that it is positive there could not have been secured the many and positive statements we herewith refer to in the detail presentation that we have prepared and enclose with this letter had not something great been accomplished and shown in the demonstrations thus far conducted.

*In this letter I have used the word "marvelous" possibly two or more times—and, I do so because I have seen the demonstration and have seen the machine torn down—I know what it will do. * * **

Yours truly,

(Signed) W. H. LOVESY.

Second P. S. Mr. Moray is much elated this morning, as yesterday in a demonstration before a Mr. Ellison, of Layton, Utah, he developed 2150 watt light and energy. This statement added at this time just to indicate the possibilities of increasing the size of the machine and the result accordingly.

(Signed) W. H. L.

———————

September 2, 1930.

Mr. T. H. Moray
2484 South 5th East
Salt Lake City, Utah

Dear Mr. Moray:

I am enclosing herewith a statement made by Mr. T. J. Yates regarding a demonstration at your home on March 16, 1929.

I have read same carefully and can say that I am interested in your device and hope something can be worked out that will be for your interest and the others connected with you.

I thank you for the opportunity you gave me on Saturday, last, of seeing a demonstration of your electrical machine. It is a wonderful accomplishment, and I hope to see you again some time soon.

Very truly yours,

(Signed) Jas. E. Ellison.

———————

The following is a telegram received by Dr. Murray O. Hayes:

Wash. 1930, Sept. 2, 9 p. m.

Murray O. Hayes:

*REPLYING TO YOUR LETTER WE HAVE COMMUNICATED WITH FLETCHER WHO ADVISES HE SAW DEMONSTRATION BUT COULD NOT GET THE DISCLOSURE OF DETECTOR. * * **

(Note the detector's weight is less than 1 ounce.)

(Dr. Fletcher has seen and had explained to him every other part of the device including the tubes).

T. H. M.

This account has now in a hurried way taken us to October, 1930. During the period to October 1, 1931, the device has been shown to hundreds including engineers and doctors of Science representing some of the largest manufacturing concerns in America and all have had to admit that the demonstrations have been wonderful and impressive. No claim is made that all those for whom the demonstrations have been made will place their okey on the device but I claim that not one has been able to find any fault or show that there have been any ear marks of this being anything but what is claimed. There are so many demonstrations that have been given that I cannot begin to cover them all but would like to mention two before closing this history.

One made in October, 1929 to Mr. A. A. Yakovlev who represented the Russian Government. There was present at that demonstration besides the above gentleman, Dr. Murray O. Hayes, the inventor, and John Magdiel. Dr. Yakovlev was first given letters covering former demonstrations and full explaination of what a demonstration would be like. The device was opened for his inspection at the close of the demonstration and he was asked if he was fully satisfied before the demonstration was discontinued.

I would like to state that all demonstrations have been the same the only difference being the type of device demonstrated as advancement has been made.

The other demonstration I would like to mention is the one given in April, 1931, to Dr. K. Vern Knudson which was preceded by a conference between Dr. Knudson and the inventor in the presence of Dr. Murray O. Hayes and Attorney Preston D. Richards. I tried to place all the cards on the table without disclosing too much of the vital eight ounces and after the demonstration and same tests had been made as described in former accounts, I opened the device for Dr. Knudson and let him examine it all. He figured that the transformer would burn up with one-fourth of the current he had seen taken from the device if ordinary A. C. current had been used and I asked him if I had told or shown him anything that was not scientifically sound, and correct and if there was any way I could have faked the demonstration and I understood him to say that no one could have faked such a demonstration and that there were no ear marks of faking at all.

Dr. Hayes later had a long conversation with Dr. Knudson in California at which time every phase of the demonstration was gone over in detail and much of the device explained to Dr. Knudson.

I have been accused of going around with a chip on my shoulder.

This is not the case but I want to get the idea over to everyone that I make no exceptions in my invitations. I invite everyone, anyone, anywhere, anytime to show one way in which such demonstrations as given above could be faked. Surely if no one can give one single explanation of how such demonstrations could be faked there are none. This thing is either all claimed for it or nothing. I fully realize some, yes, many, have a hard time to believe far more probable things than this but, gentlemen, you set your price of how much it is worth to you to furnish a sketch (simple or otherwise) of how these demonstrations could be other than all that is claimed and I will see that you are paid in full for the time spent.

This is no idle boast. This is a personal invitation to you individually, every man or woman in the world regardless of education or training and if you should not fully understand the accounts of the demonstrations I will attempt to make further explainations so that you will have ample information on which to make your drawings and explanations to expose these demonstrations.

I mean to make this invitation so clear that there will be no excuse for anyone to doubt any phase of this proposition and I make these statements without reservation of any kind. Not in any offensive way I hope as no offense is intended but because I feel I am entitled to this consideration because of the hours of time I have spent with you in going over the theory, demonstrating the device and doing everything I could to give you a chance to offer one objection or possible way these demonstrations can be other than genuine. I especially invite those whose names are mentioned in this account to accept this invitation.. Special invitation is made Dr. Murray O. Hayes, Dr. K. Vern Knudson, Mr. Thomas J. Yates, Mr. Carlos N. Gaxiola, Dr. A. A. Yakavlev, Mr. Nathaniel Baldwin, Dr. Harvey Fletcher, Dr. Carl Eyring, and any others who will be kind enough to do me the favor to make drawings, sketches, or explanations of how these demonstrations could be fixed. I have selected eight men of science as a jury. The accounts of demonstrations as given are now in your hands. As attorney for the "Device" I now rest my case. Its fate rest with you. Can you find any way in which such demonstrations could be faked?

There is nothing personal in this request. I am now out of the picture, the demonstrations now stand by themselves, I no longer defend but leave them now to stand or fall on what those who have seen the demonstraions have had to say. The Witness testimony, if you please.

Salt Lake City, Utah

October 1, 1931

To those who may be interested in the Moray Radiant Energy System, I wish to state that I have seen the system demonstrated and that light and heat are produced in liberal quantity. So far as I am able to tell the apparatus does all that is claimed for it and I see no evidence of any fraud. The amount of apparatus used is too much limited to produce all this heat and light.. The energy must come from some external source.

I examined most of the apparatus which was used consisting of condensers and coils of wire which could produce no energy themselves, but the detector (so called) a small instrument which completed the apparatus was not shown to me, neither was the principle of operation disclosed.

Most great inventions seem to be impossibilities until they are put into practical use and then they become common and seem more or less simple. I do not understand the principle of operation of this invention but I reason this way about it. Science proves that light waves, radiant heat, and radio waves are identical in character, the main differences being in the wave length and frequency. Light waves have a great range of frequency according to the color and in late years frequencies going far beyond those of light waves have been discovered and used. Heat waves have frequencies less than those of light. But what of those frequencies which are less than those of heat? We cannot say they do not exist. Our senses only respond to a few of the frequencies that have been found.

Radio waves have frequencies which are very low as compared with those of heat and light. Radio waves pass through substances which are opaque to light and heat but can be absorbed and used by the simple little crystal radio set. Radio waves are produced artificially but the sun which sends out light waves, heat waves, and a wide range of frequencies beyond the light waves may also send out powerful waves having frequencies below those of heat which may pass into the earth or through the earth and produce no effect that we know of. Other heavenly bodies also, possibly invisible, may be sending forth untold energy into space in the form of waves. Mr. Moray may be harnessing some of these waves just as the crystal set absorbs the radio waves, and possibly with apparatus almost as simple.

In view of the wonderful scientific achievements of later years one needs to think many times before he says, "impossible."

I do not believe that perpetual motion is possible in the way many have sought for it—by having a machine produce its own energy to keep it going; but where energy exists it may be possible to transform it into useful forms.

I have invited inventors of pertual motion machines to show me their plans and I would show them their mistakes, and I have done it; but this radiant energy system is different. There is a source to the energy.

(Signed) NATHANIEL BALDWIN.

At this point T. Henry Moray's brief history was discontinued, and we are assuming that this was about the time of the publication of the pamphlet. Moray also had written on the manuscript:

"A very old account gotten out by promoters taken from the Moray record, not presented as Moray would have liked it, composed and edited too much like promotion."

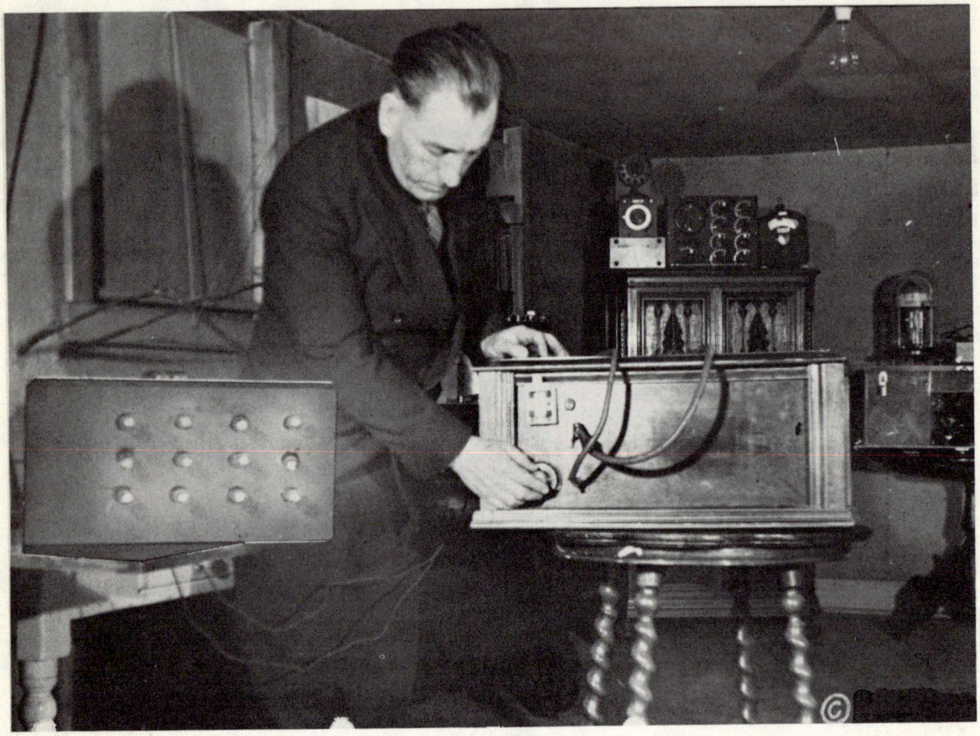

Fig. 11

Dr. Moray tuning third generation device

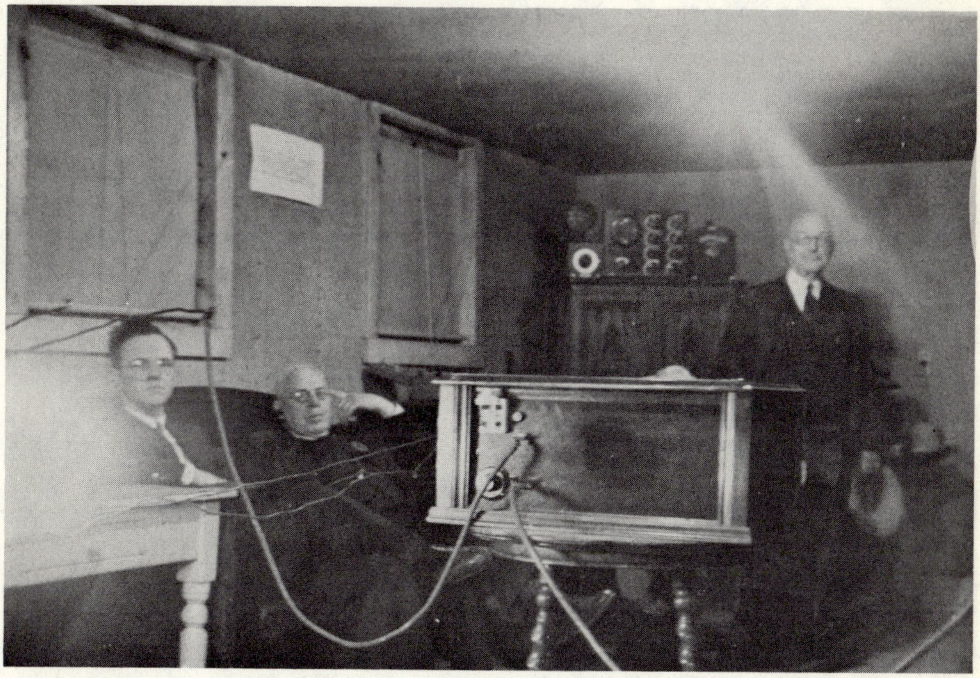

Fig. 12

Photographing often became difficult

Fig. 13

Antenna looking west to east

CHAPTER 3

"FRIENEMIES"

At the request of R.L. Judd and the Judd interests, beginning in 1925 Henry Moray had taken into his confidence consultants, first Dr. Carl Eyring and Dr. Harvey Fletcher, and then Dr. Milton Marshall, Dr. Eyring and Dr. Marshall at that time members of the BYU faculty in the physics department. Not long after this Dr. Eyring went to Bell Laboratories at the invitation of Dr. Fletcher, who was at Western Electric and then became "physical research director for the Bell Laboratories of New York City."*

Although Henry Moray was reluctant to make immediate disclosure, eventually, I contend, Carl Eyring and Harvey Fletcher learned everything Henry Moray knew about the Radiant Energy device at that time.** Step by step as he talked to Eyring, and later to Fletcher, he told these two men everything, including the fact that Milton Marshall was attempting to identify the material that he called his "Swedish Stone". When he described the action of his germanium mixture to the two physicists, they knew that the entire composition of the Swedish Stone had not been properly identified, and that Dr. Marshall was working on the project for that purpose.

Carl Eyring and Harvey Fletcher both seemed to prefer to communicate through Moray's lawyer, Judd. Dr. Marshall did correspond directly with Henry even after Henry withdrew from the Judd interest and created what became known as the Moray Products Company. At that time, Henry Moray hoped these gentlemen would still work with him as consultants. It is obvious that this belief was held by the principals of the Moray Products Company, who published the brief history we have just reviewed. Henry Moray felt that Harvey Fletcher and Carl Eyring were at least friendly observers, and he often referred others to them for recommendations on his work. Through the history of the project, however, this proved to be a mistake. It is my opinion that when Carl Eyring, Harvey Fletcher, and Henry Moray did not communicate directly, misunderstandings developed.

Dr. Fletcher insisted that Henry Moray take Carl Eyring into his confidence and show him every aspect of the project. Consequently, in an effort to please Eyring, Henry made test after test under various conditions and circumstances.

In a summary written November 19, 1925, by R. L. Judd, he tells that Henry Moray assembled the device in front of Eyring and himself. Eyring spent over an hour examining the circuits and could not find any fault in what he had seen. He

**Deseret News, Saturday, Dec. 29, 1934*

***Henry Moray's denials of this fact were only defensive in nature.*

stated that he could see that current was not obtained from batteries or hidden wires, and concluded that it was obtained by electromagnetic induction.

Henry Moray took great exception to this conclusion. In an effort to refute it, on December 21, 1925, Attorney Judd, R. L. Adams and Attorney Nebeker came to Moray's home, took the device to nearby Emigration Canyon, and experimentally set it up to show that it would still work even when the distance to the nearest powerline was too great for power induction. Then, with Dr. Moray, one morning in February of 1926, Eyring "began working on the theory and hookups, examining the device, taking it apart and putting it back together, asking questions continually, taking notes, coming up with ideas on how to apply for a patent, making drawings and sketchings, going over the details of the theory." Further quoting Henry Moray, "After spending all that afternoon with me, Dr. Carl Eyring congratulated me on what he termed my wonderful work and said what he had heard and seen was mathematically, electrically and scientifically sound and correct."

Dr. Moray explained in his note that he described the radio detector he had developed to Carl Eyring. He compared it to what was commonly known as the crystal of a crystal set. However, his detector was superior since it could drive a loud speaker without the use of a battery. He also explained what he could of the Radiant Energy Detector, although he did not allow Eyring to dismantle it, as this would have destroyed it. Similarly, he explained the function to Fletcher and later Murray O. Hayes, and even to Dr. Henry Eyring.

It is my understanding from studying the work done by Marshall and from what Henry Moray said, that Carl Eyring, Fletcher, and Murray O. Hayes were shown everything except the chemical formulas for the R. E. Moray Valve, and this was only because my father was having difficulty establishing the formulas, even though Dr. Milton Marshall was expending considerable effort to identify the Swedish stone. I contend that both Carl Eyring and Harvey Fletcher knew of this effort.

They had access to the work Dr. Marshall was doing; therefore, they knew everything even if they didn't think they did. What good did it do them to know? They still didn't understand. Solid state physics was unknown. Henry Moray had no terminology to explain to these gentlemen of science how the detector worked. Simply telling them what was in it would have been useless. Consequently, he used the most easily demonstrated device, the germanium diode, that worked on the same principle as the radio, to illustrate how he thought the Radiant Energy Detector worked. Logically, he felt that this was the only way he could make them understand what he was doing. Radio was understood; supposedly crystal sets were being investigated. Henry originally built the radio simply for that purpose - to show how he was able to pick up signals with a solid state device sufficiently strong that they would drive a loud speaker, which was something unheard of in that day. However, he was not able to reach understanding with the physicists. His circuit did not have batteries, as it was very similar to the old crystal-set circuitry.* Figure 17, sub-figure 3, shows how the original germanium valve was used and how it worked in the radio circuit.

**In investigating with the manager of one of the local chemical supply houses, I*

The earliest drawings and descriptions of the Moray Valve are found in a sworn certificate dated November 14, 1927, where he describes a germanium compound using the word "pure germanium". (Figure 14)

In 1937, in order to protect his interest in the germanium valve, he wrote an affidavit and had this affidavit witnessed (Figure 15). This affidavit refers to certain drawings and descriptions found in patent application 550611, which, according to the affidavit, is what he originally showed to Carl Eyring of BYU in his home in 1925, later to Harvey Fletcher of the Bell Laboratories in 1928, and finally to Murray O. Hayes, who was purportedly his patent attorney in 1931. I have made copies of the patent application that applied to the description found (Figure 8 and 9, patent application 550611, enclosed). (Figure 16).

The Radiant Energy Detector, even today with the advanced state of the art of semi-conductors, is considered by our research organization to be proprietary and still of great value. Parts of the certificate (Figure 15) have been removed to protect those interests. It must be noted that the detector described in Figure 15 is also superior to most germanium semi-conductors known today because of some of the doping material used. That this bi-polar device is truly a transistor was verified by examination of the complete notes by Warren Simmonds, Ph.D., of Salt Lake City.

In September, 1928, Fletcher also had examined the device and had been thoroughly briefed on the circuits and hookups. Fletcher thought the energy came from one of two sources, either an external source as claimed by the inventor, or possibly through harmonizing or synchronizing the very units comprising the equipment in such a way that chemical action results in tubes sufficient to make the demonstration. At this point Dr. Fletcher suggested working further on the device and producing one hundred of them.

This same experiment (Moray/Fletcher) was referred to by Judd in a letter to Mr. Heffeneffer of Fall River, Massachusetts: ". . . on Thursday a week ago Mr. Moray and Dr. Fletcher spent the afternoon in a detailed study of the drawing, demonstrating the circuits involved and the theory upon which the mechanism is based. This was requested by Dr. Fletcher. On Thursday last, myself, and Mr. Jensen, another interested party, went down to Moray's home. Dr. Fletcher was afforded the opportunity of studying the antenna, the ground, the respective connections with the machine. Moray then demonstrated that there was no life in any part of the machine.

"They successfully started up the device, taking about six minutes for tuning. Dr. Fletcher was given further opportunity to study its operation. Dr. Fletcher, in observing the lamps, stated they were burning far above normal, and that if they were to continue at that degree of radiancy, they would soon burn out. While observing the lights, various tests were made, such as disconnecting the antenna and the ground. The lights were then screwed out quickly, and a 575

asked him what Dad meant by "pure germanium". He said, "Your father sent it back five times for re-purification before we reached the grade he demanded." (See figure 14). Part of this certificate has been blanked out so as to protect proprietory information.

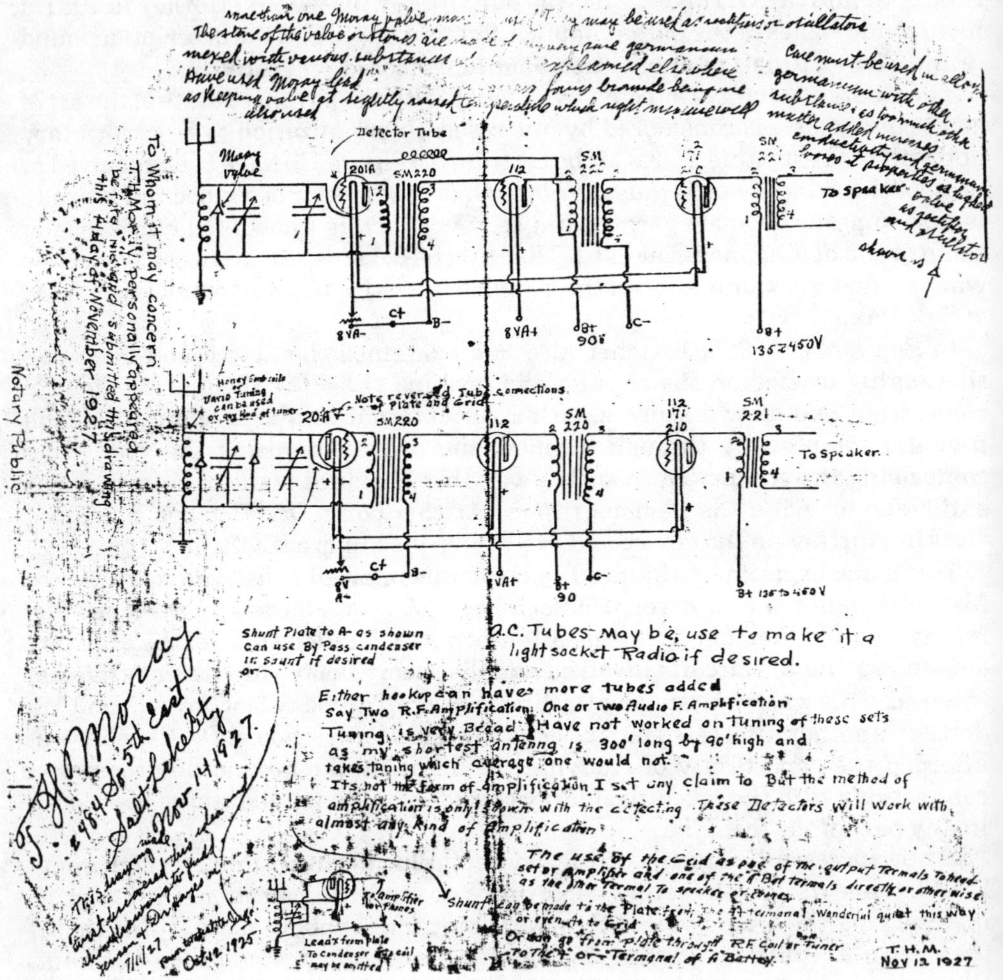

Fig. 14

Affadavit, 1927

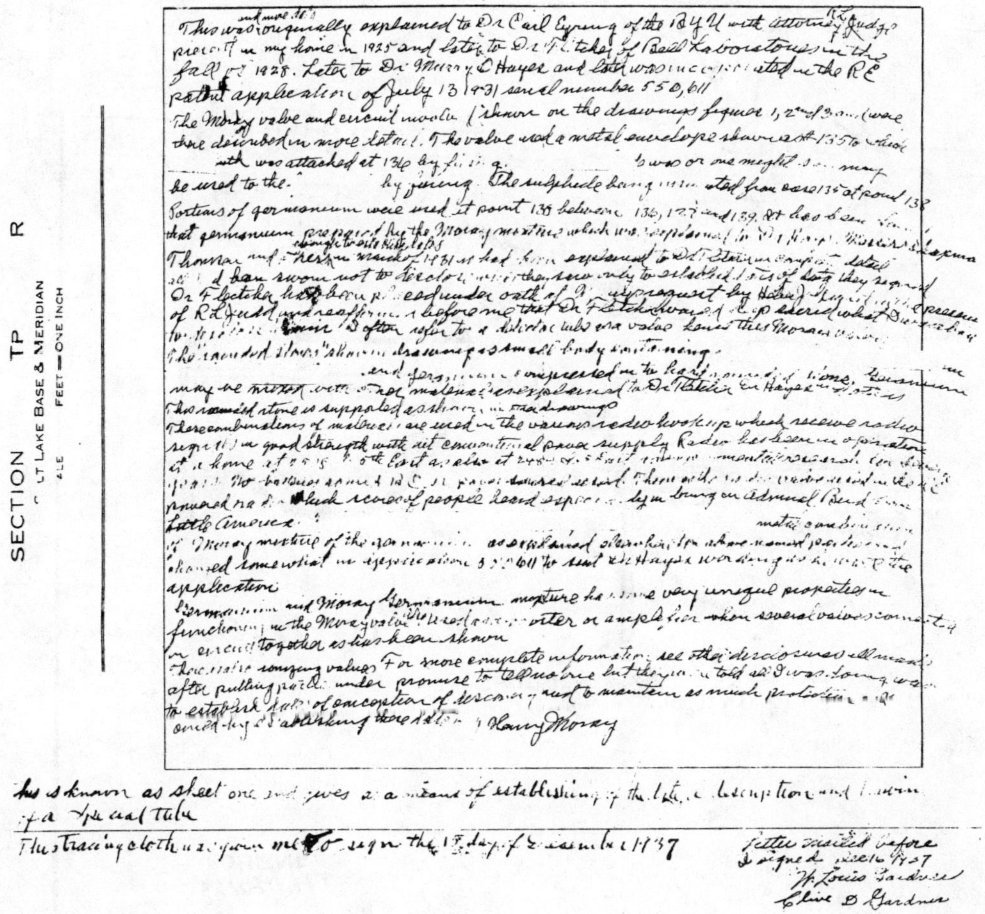
SECTION TP R

[illegible]LT LAKE BASE & MERIDIAN

[illegible]LE FEET = ONE INCH

This was originally explained to Dr Carl Eyring of the BYU with attorney Judge [illegible]
present in my home in 1925 and later to Dr Fletcher of Bell Laboratories in the
fall of 1928. Later to Dr Murray C Hayes and later was incorporated in the RE
patent application of July 13 1931 serial number 550,611
The Moray valve and circuit [illegible] (shown on the drawings figures 1, 2 and 3 [illegible]) were
there described in more detail. The valve used a metal envelope shown at 135 to which
[illegible] was attached at 136 by [illegible] [illegible] was or one might [illegible] may
be used to the [illegible] by fusing. The [illegible] being [illegible] from case 135 at point 138
Portions of germanium were used at point 138 between 136, 137 and 139. It has been [illegible]
that germanium prepared by the Moray mixture which was explained to Dr Hayes [illegible]
Thomas and [illegible] in March of 1931 as had been explained to Dr Fletcher in complete detail
[illegible] I have sworn not to disclose [illegible] they saw only to establish dates of date they signed
Dr Fletcher had been placed under oath of [illegible] my request by Hon. [illegible]
of R L Judd and reaffirmed before me that Dr Fletcher would keep secret what [illegible] presence
[illegible] I often refer to a detector tube or a valve known this Moray [illegible]
The [illegible] shown in drawings as small body containing [illegible]
[illegible] and germanium [illegible] in the [illegible] stone. Germanium
may be mixed with other metals [illegible] explained to Dr Fletcher Dr Hayes [illegible]
This [illegible] stone is supported as shown in the drawings
These combinations of metals [illegible] are used in the various radio hookups which receive radio
signals in good strength with out conventional power supply. Radio has been in operation
at a home at [illegible] East as also at [illegible] East [illegible] months [illegible]
[illegible] No batteries [illegible] power source [illegible] with no [illegible]
[illegible] scores of people heard [illegible] Admiral Byrd from
Little America
[illegible] metal combinations
of Moray mixture of the germanium [illegible] as explained [illegible] above named [illegible]
changed somewhat in application 550,611 to suit Dr Hayes wording [illegible] the
application
Germanium and Moray Germanium mixture has some very unique properties in
functioning in the Moray valve [illegible] rectifier or amplifier when several valves connected
in circuit together as has been shown
[illegible] ionizing values. For more complete information see other disclosures [illegible]
after pulling [illegible] under promise to tell no one but they [illegible] told [illegible] I was doing [illegible]
to establish date of conception of discovery and to maintain as much protection [illegible]
[illegible] establishing these dates [illegible] Henry Moray

[illegible] is known as sheet one and gives as a means of establishing of the date [illegible] description and drawing
of a special tube

This tracing cloth was given me to sign the 1st day of December 1937

Letter mailed before
I signed Dec 16 1937
W. Louis Gardner
Clive D Gardner

Fig. 15

Written description of Moray Valve (MV) for patent

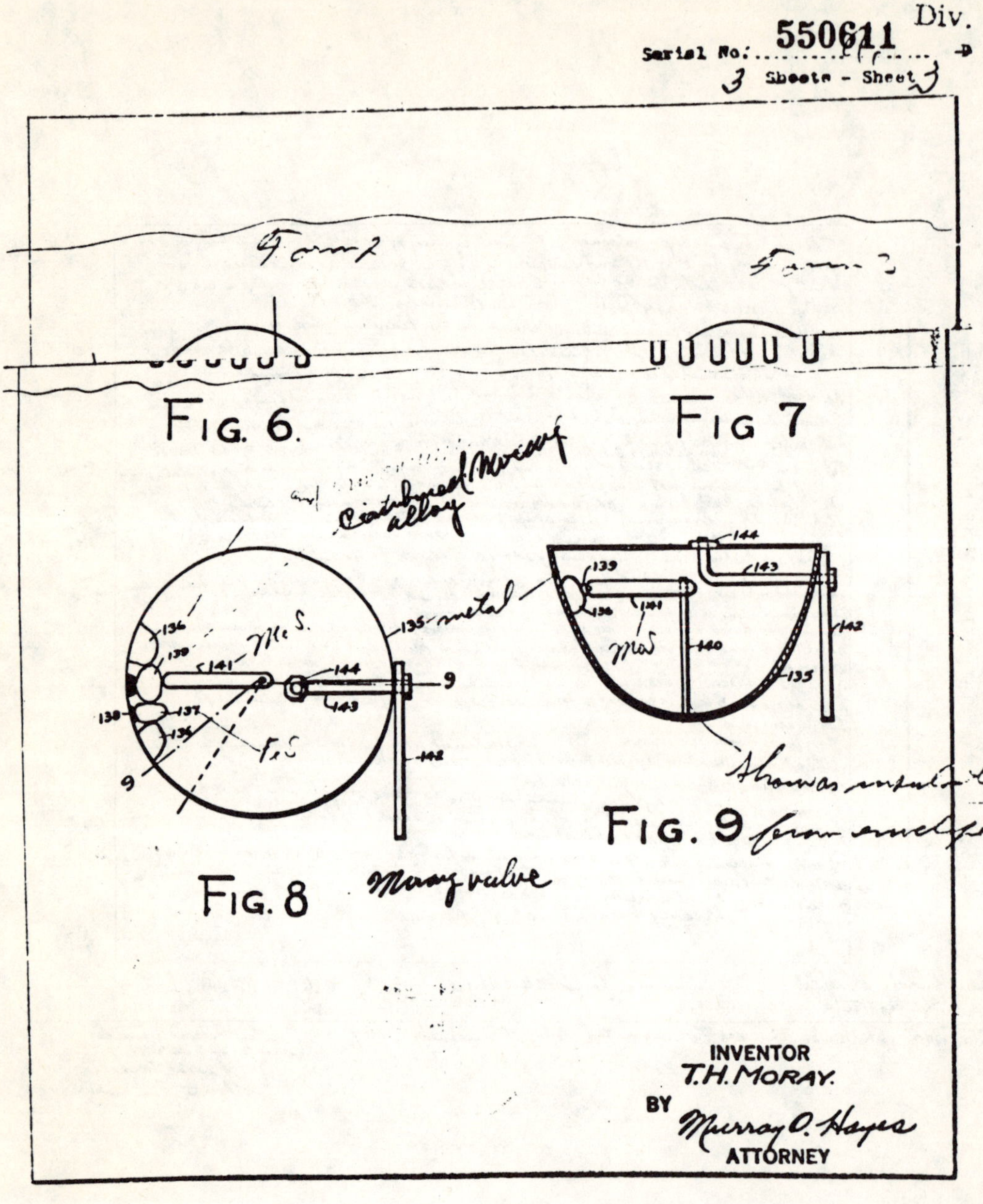

Fig. 16

Patent drawing

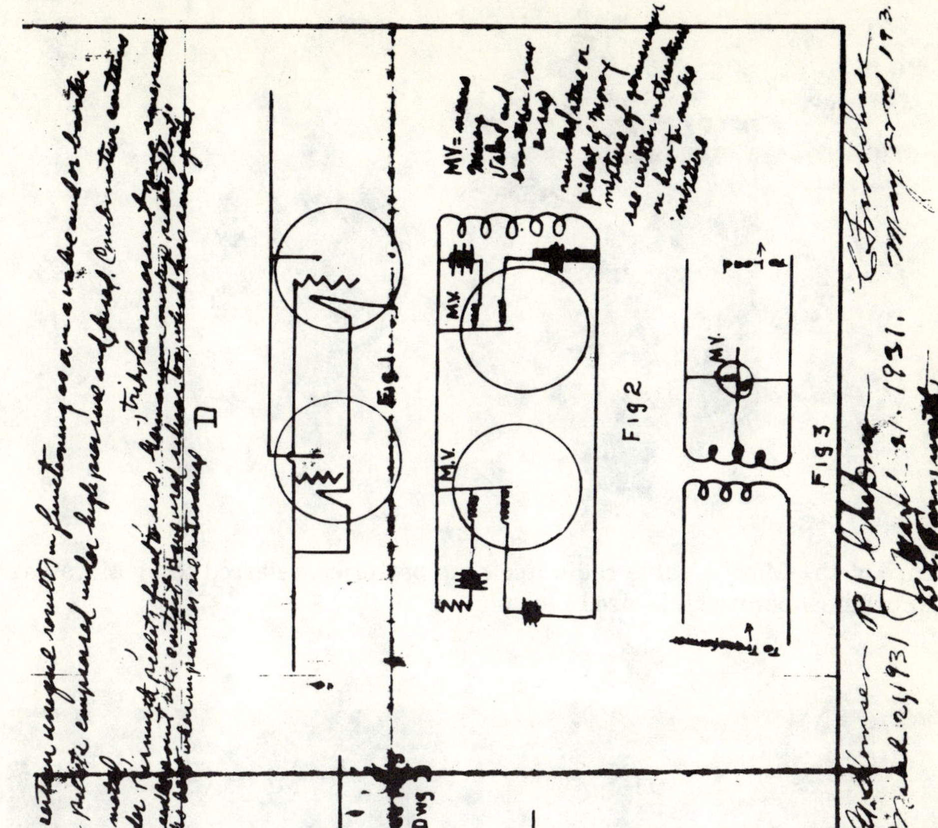

Fig. 17

Drawing, March 21, 1931

Picture of the Moray Valve radio itself, no batteries, referred to by Moray as pocket radio: operated full-sized speaker

Moray Valve radio powered by Radiant Energy.

Fig. 18

watt flat iron attached. This was heated to sizzling point in about 5 minutes. Then the wires were disconnected and the lights again put on. After the lights had been burning for some little time, for some reason, then unknown, they flickered a few minutes and then went out. It was determined after some investigation, that the cat's whisker on the detector had become disengaged, and therefore the detector had ceased to operate. Encounters of difficulties only proved that Dr. Moray had not pre-planned the demonstration, and that it was pure investigation. According to Mr. Judd, Dr. Fletcher agreed that the machine was what Dr. Moray contended it to be. He said that Fletcher commented with regard to 'battery action kept alive by some form of energy from the atmosphere' and then said, 'From this demonstration it is established, I should say absolutely, that we have something entirely new and very wonderful and that there is no possibility for doubt in any phase of this proposition.' "

Judd's letter concluded, "Dr. Fletcher left for New York by automobile yesterday morning. He told me that following the time he had to sit down and think out this matter a little and he would write us and try to make some suggestions for further development."

The argument then became no longer one of induction, but of how long the device would endure. If it were some sort of battery action, could it last? So it was arranged that a time test would be made. The details are in the letter written by Robert L. Judd to Dr. Harvey Fletcher of the Bell Telephone Laboratories, where Mr. Judd outlines the procedure used when Dr. Murray Hayes and Mr. Jensen put the device through a series of time tests running from October 1 to October 6 or 7, 1928.

Soon after, Henry Moray and Carl Eyring had a date at BYU, and Eyring asked that particular tests be conducted to demonstrate specific levels of wattage. Henry Moray pointed out that this was impossible, because with the amount of wattage that Carl Eyring insisted upon, the amperage would burn out his device. Carl insisted that Henry Moray could demonstrate this without overheating the detector. As $E=IR$ is the most basic formula in electronics, Henry had to insist that Carl was wrong. Henry put the needed formulas on the blackboard and insisted that either Carl Eyring would admit that they were correct, or else Henry would call in physicists, including Dr. Marshall, from the BYU physics department. Carl Eyring insisted that they were not correct, but finally, rather than have other physicists involved, he admitted in front of Mr. Judd that Henry Moray's formulas were, indeed, correct. As a result, Henry Moray felt he had embarrassed Eyring in front of R. L. Judd. (And Henry Moray always wondered if Eyring didn't blame him for the fact that he never became president of the Brigham Young University.)

When Carl Eyring and Harvey Fletcher began to work as consultants on the project, R. L. Judd had sworn them to secrecy. Henry Moray believed that Harvey Fletcher had promised to provide help with the project because of letters delivered to him from Judd, his attorney. Early indications of trouble developed in 1928; however, Henry seems to have ignored most of them. A letter (quoted below) to R. L. Fletcher, passed to Henry through a third party, is important here. It gives, first, a notation written by Henry Moray which indicates that a disagreement was developing with Carl Eyring; second, it gives us an analysis of how Harvey Fletcher viewed Henry Moray's discovery; third, it gives a means by which Harvey Fletcher is suggesting that the project be handled.

Mr. Judd, at this time, did not pursue the points the letter raised. Henry was still depending on Judd, as his own attorney and as the representative of what was known as the "Judd syndicate," to guide him in business activities such as these.

There are several mistakes included in the copy of the letter as Dr. Ostler's secretary typed it and sent it along with a personal note from Ostler written on a prescription pad.

"Dr. D.E. Ostler
Eureka, Utah

"(Copy of the letter written by Dr. Harvey Fletcher of the Western Electric Company under the date of Oct. 6, 1928 to Mr. Robert L. Judd of Salt Lake City. The letter was written in New York City.).

"Dear Judd:

"We had a very pleasant return trip, arriving here in about six and a half days. The constant driving made me somewhat tired but after two days rest I feel fine. One can get a very good conception of what our country is like by driving across it in an automobile.

"Now regarding the experiments Mr. Moray showed us, I will say at the outset that I am just as puzzled as ever. I can give no satisfactory explanation of the result. If I saw all the parts that enter into the production of the light I would certainly agree with Mr. Moray that either the tubes or the rectifier or the coil had some very remarkable properties. As a scientist I should like to see them investigated in some physical laboratory which is equipped to do such work. If Mr. Moray's statement that the tubes have a capacity of a farad is even approximately true, the tubes alone have a great scientific value.

"The evidence as presented seems to favor Mr. Moray's explanation of where the energy came from. However, because it is so contrary to all previous notions about electrical sources and also because Mr. Moray was unable or unwilling to state how the various parts functioned, I am still of the opinion that all of us, including probably Mr. Moray, have overlooked something which will explain the lighting of the light in an orthodox way.

"There are certain facts which became evident to me as I saw the experiments:

"(1) There is considerable energy drawn from somewhere. Apparently you have satisfied yourself that it is not from other power stations in the city. Then it must be in the set itself. This looks improbable although not impossible. Some careful experiments in a laboratory would settle this points.

"(2) The energy is transferred from a high impedence circuit to a low impedence circuit by means of a high frequency current. The high frequency is probably produced by an oscillatory circuit in the system and I think tests would reveal that the frequency of oscillation was entirely controlled by the constants of the circuit and not by outside influences. Any attempt to obtain current or voltage readings on the high impedence side of the circuit by ordinary meters would probably result in failure. This is confirmed by Mr. Moray's experience.

"(3) If the source of energy is within the system, by redesigning the system the same performance can be obtained without the use of the antennae.

"(4) If the rectifier has only the function Mr. Moray claimed for it, then a substitute can easily be found which is much more stable and reliable.

"Assuming Mr. Moray is correct in his explanation, in my opinion it would be many years before he would be able to perfect his device by working all alone by the cut-and-try methods that he must necessarily use. Progress is not made in these days by lone workers. There are so many phases to such a problem that it requires the cooperation of specialists to answer satisfactorily the different phases of the problems. Unless Mr. Moray changes his attitude it seems to be hopeless to expect any progress whether he is right or wrong. He expects everybody to trust him and give him support but still he will trust nobody. When he will take into his confidence such fine men as Marshall and Eyring to such an extent that they can duplicate his apparatus I really think something good will come out of it, probably in quite a different way than he now expects."

"Sincerely yours,
Signed (Harvey Fletcher)*"

In the foregoing letter, Dr. Harvey Fletcher agreed that there was evidence of considerable energy being drawn from somewhere. If the energy was within the device itself, as he supposed in the third paragraph, then the discovery was even greater than Henry Moray's claims that the energy came from the universe. Both Harvey Fletcher and Carl Eyring had been given the opportunity to test each one of the components, including the detector, to determine whether there was any energy in the components. Fletcher implied that there was no known souce of energy outside of the set itself "if it is not induction." Referring to the detector as the "rectifier" he discussed how easily this material could be found; in my opinion, showing he knew that the detector had originally been found by Henry Moray and contained substances that he was not familiar with. No one had properly identified this substance at the time. Henry Moray had explained to them everything he knew about it, and they were asking questions which were beyond his capability to answer.

Fletcher's emphasis then goes to his claim that Henry Moray was not taking Marshall and Eyring into his confidence in order for them to duplicate the Radiant Energy device.

*****_Dr. Moray's notes concerning this letter are handwritten on the copy we have - "He did see all and handled all but one part which weighed one ounce."_

This only amplifies my belief that Eyring and Marshall were working together and, from his paragraph 4, establishes that they knew as much as Henry Moray about the Swedish stone. If Carl Eyring was unable to duplicate Henry Moray's work, it was simply because he did not ask to be allowed to do so. At what point this letter was delivered to Henry Moray, I do not know.

Communication broke down between Henry and the others. The decision became whom to listen to and whom to trust. His friend of nearly 20 years, R. L. Judd, withdrew because of his interests with powerful political people who had been offended when Henry Moray created the Moray Products Company. Of course, the final separation of Henry Moray and the Judd interests, although they were not openly hostile, left many people with hard feelings. Still, with the formation of the Moray Products Company, Henry hoped to keep a good relationship with Carl Eyring and Harvey Fletcher. This, however, did not follow. As he had trouble with patent applications and as his problems with the Moray Products Company multiplied, Henry heard rumors that disturbed him as to Eyring's and Fletcher's intentions. At the time, Henry Moray was alienated from Carl Eyring and consequently from Henry Eyring. He was also alienated from Harvey Fletcher and other scientists who depended upon Fletcher and the Eyrings for explanations of what Henry Moray was doing.

I am convinced that the Eyrings and Harvey Fletcher were also alienated because of the political influence of Judd interests, even though Henry Moray did not at first recognize it. Because my father did not clearly understand the circumstances and these men's attitude, he continued to send people to Dr. Fletcher for references. Instead of asking Henry not to use his name, Fletcher gave bad references to those people who wrote him about my father.

Years later, after the Moray Products Company had been placed in receivership, Henry Moray found a telegram (Figure 19) from the American Industries and Management Corporation of New York, addressed to Murray O. Hayes of the Moray Products Company, stating that they had talked with Dr. Fletcher and were satisfied that "the thing is right for financing." For some reason the deal fell through. Hayes had never informed Henry Moray about this company's interest in him. This telegram along with Millikan's letter gives the first indication of what Fletcher was saying at the time. It also gives an indication of how Henry Moray was isolated from those around him. Recent discussions between Richard Moray and Harvey Fletcher indicate that Fletcher had known nothing of this telegram.

In an attempt to get further scientific substantiation, Mr. William Lovesy wrote Dr. Robert Millikan of the California Institute of Technology in an effort to get him to go over the device. Dr. Millikan answered that he was not interested in investigating the device beyond what Dr. Fletcher and Dr. Eyring had done. Then he added that he was not interested because he was always suspicious of a person who is trying to conceal things from a group of scientists. Henry Moray's comment on the side of this letter was "Moray has never tried to conceal anything." (See Figure 20). Just when this letter came into Henry Moray's hand from Lovesy, I do not know. However, it shows that even then Fletcher and Eyring contended they had not been shown everything, although

PATRONS ARE REQUESTED TO FAVOR THE COMPANY BY CRITICISM AND SUGGESTION CONCERNING ITS SERVICE

1220S

CLASS OF SERVICE

This is a full-rate Telegram or Cablegram unless its deferred character is indicated by a suitable sign above or preceding the address.

WESTERN UNION

NEWCOMB CARLTON, PRESIDENT — J. C. WILLEVER, FIRST VICE-PRESIDENT

SIGNS

DL	= Day Letter
NM	= Night Message
NL	= Night Letter
LCO	= Deferred Cable
NLT	= Cable Letter
WLT	= Week-End Letter

The filing time as shown in the date line on full-rate telegrams and day letters, and the time of receipt at destination as shown on all messages, is STANDARD TIME.

Received at 278 Main St., Clift Building, Salt Lake City, Utah

NA286 117 NL=NEWYORK NY 2 — 1930 SEP 2 PM 8 26

MURRAY O HAYES=

405 ATLAS BUILDING SALTLAKECITY UTAH=

REPLYING TO YOUR LETTER WE HAVE COMMUNICATED WITH FLETCHER WHO ADVISES HE SAW DEMONSTRATION BUT COULD NOT GET THE DISCLOSURE OF DETECTOR WE ARE NOT CONNECTED WITH ANY POWER OR ELECTRIC TRUST BUT INDEPENDENTLY REPRESENT GROUP OF PEOPLE WHO PLEDGE THEMSELVES TO FINANCE WORTHWHILE SCIENTIFIC DEVELOPMENTS WE DO NOT HAVE TO SELL STOCK IF WE ARE SATISFIED THAT THE THING IS RIGHT FINANCING CAN BE ARRANGED BETWEEN EIGHT OR NINE PEOPLE THE ONLY LOGICAL THING TO DO WOULD BE FOR YOU AND DR MORAY TO COME TO NEWYORK AT ONCE AND WE WILL PLACE AT HIS DISPOSAL PROVIDED WE ARE SATISFIED ALL FINANCES NECESSARY TO COMPLETE INVENTION TOGETHER WITH ALL OTHER SCIENTIFIC FACILITIES THAT HE MIGHT REQUIRE=

AMERICAN INDUSTRIES AND MANAGEMENT CORP.

For some reason the deal fell through
Dr. Hayes never informed Moray as to what took place.

THE QUICKEST, SUREST AND SAFEST WAY TO SEND MONEY IS BY TELEGRAPH OR CABLE

Fig. 19

Telegram to M.O. Hayes

CALIFORNIA INSTITUTE OF TECHNOLOGY
PASADENA

NORMAN BRIDGE LABORATORY OF PHYSICS

November 5, 1929

Mr. W. H. Lovesy,
c/o Utah Oil Refining Co.,
Salt Lake City, Utah.

Dear Mr. Lovesy:

Replying to your letter of October 26th, my former letter stated my position completely with respect to your proposal. If Mr. Fletcher and Dr. Eyring* have the time and the inclination to investigate this matter I am glad to tell you that they are competent people to do it, and I should have great confidence in their findings. It is utterly impossible for me to take the time from the important matters I am engaged in to investigate it myself. In addition to that, it does not interest me as stated for the reason that I am always suspicious of a person who is trying to conceal things from a group of scientists.

Moray has never tried to conceal anything

Very truly yours,

Robert A. Millikan

RAM:IH

*Dr Carl Eyring

Fig. 20

Letter from Dr. Robert Millikan to W.H. Lovesy

Henry Moray believed he had withheld nothing.

If Eyring and Fletcher had been as supportive as it may appear from the cable of September 2, 1930, why was Robert Millikan suspicious of my father? It would appear that from this point Fletcher and Eyring had no further contact with Henry Moray and, more and more, Fletcher refused to be involved or even to discuss the subject with anyone.

Many people continued to write for recommendations on my father's work, particularly to Harvey Fletcher because of his position with the Bell Laboratories. Negative reactions began to filter back. On March 30, 1933, Dr. C. R. Benzel, a chiropractor, wrote Fletcher in an effort to verify Henry Moray's claims. Fletcher wrote back on April 17 (See Figure 21), saying, "I did not have an opportunity to go over his apparatus thoroughly." Henry Moray's reaction to the letter was to write a memo, "Plain untrue, see the letter from Attorney Judd, Mr. E.G. Jensen." Fletcher had seen the circuit and had seen it work. He had had the opportunity of checking the circuit against the diagram.

In May of 1933, a Mr. Alvin Todd of San Francisco wrote Fletcher, and Fletcher replied: "In thinking over the amount of work necessary to make a complete report of what I know about Dr. Moray's device, what I know about our other individuals and companies who have attempted to commercialize it and also my professional advice to anyone who is expecting to finance a company to exploit the device. I would conclude it would require about $200 worth of my time." Apparently, Dr. Fletcher at this point simply does not want to be further involved. To the best of my knowledge, Dr. Fletcher was in no way involved in any of the companies or individuals trying to exploit the device.

Then in 1937, K.K. Steffenson wrote Fletcher from Los Angeles. What Bell Labs wrote (Figure 22), with memos written by Dr. Moray when he saw this letter, is enclosed. Dr. Fletcher's reply (Figure 23) shows his disinterest in the Moray discovery. Henry Moray's memos show that there were definitely two sides to what took place.

In 1956 Dr. Moray first heard of the recommendation Fletcher made at the time he was with the College of Physics and Engineering at Brigham Young University in Provo. Fletcher stated, "I have recommended to him and to his sponsors that the only way he will ever get recognition for his Radiant Energy machine is to have a respectable engineering firm build it, and prove that it will work. Since he has failed to do this, I have no confidence in both his theories and his machine."

On June 5, 1957, again Dr. Fletcher made a claim that he had gotten a corporation to agree to build the Moray device (Figure 24).

In all fairness to Dr. Fletcher, we can never know for certain what actually happened. We only know that, because of the offenses he suspected, Henry Moray considered Dr. Fletcher to be an enemy.

Recently an engineer with whom I have been associated on a number of projects, Glenn Foster, had an opportunity to talk to Dr. Harvey Fletcher, Sr. This engineer mistakenly thought that Fletcher had been a great friend of Henry Moray's and spoke enthusiastically of Henry Moray's discovery. Foster's statement is printed in its entirety (Figure 26).

April 17, 1933

Dr. C. R. Benzel

708 Ninth Ave.,

Greeley, Colorado.

Dear Dr. Benzel:

It is true that I did see the device which Dr. Moray has invented, and the fact that I did has caused me considerable inconvenience as I have received letters of inquiry from all over the United States. I did not have an opportunity to go over his apparatus thoroughly, so I cannot say what is is that is producing the effect.

In answer to your first question, I was not able to determine the nature of the current. I judged from its effects that it was a high frequency current. In answer to your second question, I do not know what were the parts used. Dr. Moray showed me a circuit but he seemed to be unable to know enough about electrical circuits to draw a real circuit diagram. The one which he showed me as inoperative. In answer to your last question, at first I thought Mr. Moray was sincere and was simply self-deceiving in thinking he had the effect. However, recent developments have led me to question his sincerity.

Sincerely yours,

Harvey Fletcher
Acoustical Research Director

Untrue

(The original is in my files at 708 9th Ave.,

/s/ C.R.Bensel

Fig. 21

Letter from Dr. Harvey Fletcher to C.R. Benzil

BELL TELEPHONE LABORATORIES
INCORPORATED
463 WEST STREET NEW YORK
CHELSEA 3-1000

August 2, 1937
IN REPLY REFER TO
JBK-328-GZ
REPLYING TO

MR. K. K. STEFFENSEN
Hotel Cecil
Seventh & Main Streets
Los Angeles, Cal.

Dear Mr. Steffensen:

Dr. Fletcher is in Europe at present. In his absence I am replying to your letter of July 30 regarding the apparatus which Dr. Moray has developed.

Looking through some correspondence which Dr. Fletcher had several years ago concerning the Moray apparatus I believe the following represents his experiences and judgment in the matter. The demonstration was so contrary to his expectations that he was very much interested. He was not allowed to examine all the parts of the device and in consequence had no idea how it operated. Despite his inability to determine how the apparatus functioned, Dr. Fletcher felt that there was very little chance of the device operating in the manner claimed for it by the inventor. However, he stated that even a small chance of success warranted a complete investigation of the device. He suggested that if Dr. Moray wished to have the device accepted by the scientific world it would be necessary for him to explain its operation to a reputable scientist and have this scientist duplicate Dr. Moray's performance.

Not true see other letters

does Dr. Fletcher have so little confidence in his scientific ability

I hope these comments may be of some assistance to you.

Very truly yours,

J. B. Kelly

J. B. Kelly.

Fig. 22

Letter to K.K. Steffenson from Bell Labs

BELL TELEPHONE LABORATORIES
INCORPORATED
463 WEST STREET NEW YORK

CHELSEA 3-1000

September 22, 1937
IN REPLY REFER TO
HF - GZ
REPLYING TO

MR. K. K. STEFFENSEN
211 Horizon Avenue
Venice, Cal.

Dear Mr. Steffensen:

I have just returned from Europe and have read some of the letters which you have sent to the Laboratories. It is true that I do know something about the Moray patent but to recount its history from the many letters which I have received would be a laborious task. If you wish this information and will mail me a check for $100 I will be glad to get it for you.

Very truly yours,

Harvey Fletcher

Physical Research Director.

Fig. 23

Letter from Dr. Harvey Fletcher to K.K. Steffenson

Fletcher may indeed have tried to invite Moray to Bell Laboratories. But I am sure that if the proper invitations had been made directly to him, Henry would have responded in some way. It would have been typical for Moray to have said, "I would be glad to go to Bell Laboratories if they will do so and so." However, no such correspondence exists in Moray's files.

The final blow to Dr. Moray was in 1959 when he received a letter from Ernest Wilkinson, then President of Brigham Young University (Figure 27). This letter was a great shock to Dr. Moray since he had not at the time made any inquiries to either Dr. Fletcher or Dr. Wilkinson, or to Brigham Young University, that would evoke such a letter. The letter again implied that Dr. Fletcher had had no opportunity to verify Moray's device, adding that "until you're willing to have your ideas tested and verified in an acceptable manner I fear there is nothing we can do to assist you."

Wilkinson was implying that Henry Moray had not made a complete disclosure to Fletcher. I contend, knowing Henry Moray as I did and from my investigation of the record, that Fletcher had known everything. Wilkinson made an error when he asserted that an inventive scientist makes disclosures. Indiscriminate disclosure such as publishing work in journals or presenting it to "juries of qualified scientists" becomes a bar to patentability. Only when a scientist becomes a consultant working on a project can disclosure be made to him—as was made to Fletcher, Eyring, Marshall and the patent Attorney Hayes—without making the work public property.

Henry Moray, knowing that he really had made complete disclosures to Fletcher, could never understand why Ernest Wilkinson said what he did. Finally, we have the sworn statement of Glen Foster as to what Fletcher actually said at a later time (Figure 26). I leave it to the reader to examine the material carefully and draw his own conclusions.

Henry feared that Fletcher had used the information given him for his own purposes. By 1948 it was Henry Moray's understanding that Fletcher had been in the department in Bell Laboratories that developed the transistor. He could not dispel the feeling that the worst violation of confidence had taken place.

To quote from page 133 of the original "Sea of Energy" Henry Moray says:

"The Moray files, signed drawings, records and documents, together with scores of witnesses who heard the Moray radio operated by the Moray germanium valve, all prove that Moray discovered and experimentally operated the device now known as the transistor some twenty years prior to the time the Bell Laboratories did so. The records show this discovery was disclosed as early as 1925 to Dr. Carl Eyring who was later with the Bell Laboratories and a cousin of Dr. Fletcher. Also that in 1928 it was shown to Dr. Harvey Fletcher who became the head of the division of the Bell Laboratories which came out with the transistor in 1948.

"It is the opinion of the editors that it is certainly strange that two employees of Bell Laboratories came out with a germanium mixture that prototyped the Moray valve. We doubt that the Bell Laboratories know the full story of this discovery as they, being an honorable company, would never have been a party to such an unethical action."

BRIGHAM YOUNG UNIVERSITY
PROVO, UTAH

June 5, 1957

COLLEGE OF PHYSICAL & ENGINEERING SCIENCES

Mr. John Hand
General Delivery
Calgary, Alta, Canada

Dear Mr. Hand:

I will have nothing further to do with Moray or his invention. I went to the trouble of arranging for an electrical concern in New York to build his apparatus and to try it out. The reason it was never done was that on the day that Moray was supposed to arrive in New York he sent a telegram saying he would not come. From that day I have felt that he is trying to deceive the public. I have no way of deciding whether or not his invention is genuine simply because he refuses to let anybody see it or build one like it.

This N.Y. statement 100% untrue in every respect

Sincerely yours,

Harvey Fletcher

Harvey Fletcher, Dean

HF/dg

Dr Fletcher had the device dismantled for him part by part

Fig. 24

Letter to John Hand from Dr. Harvey Fletcher

BRIGHAM YOUNG UNIVERSITY

PROVO, UTAH

COLLEGE OF PHYSICAL & ENGINEERING SCIENCES

October 22, 1956

Dear Mr. Schiess:

It is true that I witnessed a demonstration by Mr. Moray in Salt Lake some 15 or 20 years ago and it has been a source of annoyance to me ever since.

I have recommended to him and to his sponsors that the only way he will ever get recognition for his radiant energy machine is to have a respectable engineering firm build it and prove that it will work. Since he has failed to do this, I have lost confidence in both his theories and his machine.

Sincerely yours,

Harvey Fletcher

Harvey Fletcher

HF:dr

why

not true

he never recommended any thing to me

However where would Moray's rights be then

ie Section 4886 of patent code when the device is not patentable

We have letters showing that Dr Fletcher had the device dismantled part by part for his examination and he OK what he saw as sound and correct

Fig. 25

Letter to Mr. Schiess from Dr. Harvey Fletcher

October 31, 1973

To Whom It May Concern:

Early this spring (1973) Harvey Fletcher, Jr had been invited by members of the Systems Engineering section on one of the weapon systems at Hill Air Force Base, Utah, to discuss details of a mathematical analysis relating to the system. Harvey, Jr at that time was on the Brigham Young University faculty.

Harvey, Jr brought his father, Dr Harvey Fletcher, with him. Knowing of his prominence in the scientific and engineering world I took the opportunity to visit with Dr Fletcher in an empty conference room while the technical meeting with Harvey, Jr took place in an office down the hall. I spent nearly 20 minutes alone with Dr Fletcher before we were joined by Harvey, Jr and two other government engineers, one of whom was Don Jones. All three of us government engineers had some knowledge of T. Henry Moray's scientific discoveries and were curious as to Dr Fletcher's knowledge of these same discoveries.

To my knowledge Dr Fletcher was not aware that the three of us had to some degree investigated the Moray discoveries. Harvey, Jr had learned on a previous occasion that we were interested in the Moray energy findings. I directed a number of questions to Dr Fletcher which were designed to remind him of his earlier work including his position with the Bell Labs.

Dr Fletcher described some of his experiences with Edison and Millikan and stated that he, Fletcher, was actually responsible for the so-called Millikan oil-drop experiment although Millikan did collaborate with him. Circumstances were such that Millikan received the credit for the famous measurement of the charge on an electron. Harvey, Jr joined in the conversation impulsively for a moment in support of his father's activities in science.

After this preliminary discussion and establishment of a friendly rapport, I asked several other questions to determine if Dr Fletcher had had any contact with T. Henry Moray. I did this in a manner such that he wouldn't take offense at the questions or feel that I was biased in any way. I had heard from others that Dr Fletcher was sensitive about the subject.

Dr Fletcher stated that indeed he had known and talked with Henry Moray, had observed the energy device functioning, and further observed that there was a marked interest in technical circles about the Moray energy discovery along with some skepticism. He remarked that some, including himself, desired to test the device under laboratory conditions at the Bell Labs in New York to determine its validity and functional characteristics. Dr Fletcher went on to state that Mr Moray had been invited by him to come back to the Bell Laboratories and that Mr Moray had agreed but did not arrive in New York for reasons unknown to Dr Fletcher.

Fig. 26

Letter from A.G. Foster

Page Two 31 Oct 73.

During the entire conversation Dr Fletcher talked with lively candor and interest about both Mr Moray and Dr Millikan. I have concluded, therefore, that Dr Harvey Fletcher was sufficiently aware of the Moray discoveries and in particular the energy device known as "Radiant Energy" to promote an interest in it at the Bell Laboratories. To my memory the time period involved was the 1930's but no specific date was mentioned during the conversation with Dr Fletcher.

Arthur G. Foster

ARTHUR G. FOSTER
Professional Engineer (EE)
Utah 2222
Ohio 19285

Subscribed and sworn to before me this 12th day of November 1973.

Juanita L. Kukakeki notary public

Fig. 26

Letter from A.G. Foster *(continued)*

BRIGHAM YOUNG UNIVERSITY

PROVO, UTAH

ERNEST L. WILKINSON, PRESIDENT

JOSEPH T. BENTLEY, ADMINISTRATIVE ASSISTANT
CLYDE D. SANDGREN, GENERAL COUNSEL AND SECRETARY

WILLIAM E. BERRETT, VICE PRESIDENT
EARL C. CROCKETT, VICE PRESIDENT
HARVEY L. TAYLOR, VICE PRESIDENT

March 4, 1959

Dr. Henry T. Moray
25 South 5th East
Salt Lake City, Utah

Dear Dr. Moray:

Our Dr. Harvey Fletcher indicates that he has had previous contact with you and that he went to some inconvenience at one time to attempt to verify your quite astounding claims. He informs me that generally a new discovery or idea is verified by disclosure to other qualified scientists so that they may, by following the originator's directions, repeat his discovery. This method has been used in the scientific world since the time of Galileo.

Scientists generally are quite anxious to disclose their work so that it may be verified and developed. Our patent laws provide protection for patentable new inventions, and publication establishes prior right to new ideas.

Dr. Fletcher indicates that you have consistently refused to make any such disclosures in accordance with accepted scientific procedure. Each of your demonstrations has had an element of the mysterious about it, which makes it impossible to completely determine the genuineness of your claimed discoveries.

Until you are willing to have your ideas tested and verified in an acceptable manner, I fear there is nothing we can do to assist you. If and when you are willing to publish your work in accepted scientific journals, and have your discoveries evaluated by an impartial jury of qualified scientists, we shall be most happy to assist you in any way we can.

Sincerely yours,

Ernest L Wilkinson

Ernest L. Wilkinson
President

ELW:blb

Fig. 27

Letter from Ernest Wilkinson

It is hard to believe that Fletcher did not recognize the similarity of Henry Moray's bipolar germanium diode and the work conducted by William Shockley, John Bardeen, and Walter H. Brattain. It is inconceivable to me that Fletcher could observe the work going on at the Bell Laboratories (as, Henry Moray understood, the head of the research department) and not see the similarity in the use of germanium to what Henry had done. Even if the discoverers of the transistor, John Bardeen and Walter H. Brattain, received no assistance from Fletcher at all, why did not Fletcher point out the similarity of the work and at least give Henry Moray credit for the work he had done? This same situation has developed in my research. An engineer who signed confidential disclosure agreements with Cosray Research Institute before Henry Moray died, is working with the Eyring Research Institute - which has shown great interest in Henry Moray's work in what is now called "Direct Energy Conversion Systems." This involves using radioactive material in conjunction with the quartz junction. Originally, the engineer I mentioned agreed that the Eyring Institute would give credit to Dr. Moray for his work in this area. However, from the information I have been able to obtain, the Institute is not giving him credit. I am not contending they are stealing his work, I am only contending that they are paralleling the work done in 1929-'30. The fact that they are using a different radioactive material from what Henry Moray used is of no importance.

One may well ask what damage was done.

The Eyring family is an old closely-knit pioneer family. Dr. Carl Eyring is a cousin of Dr. Henry Eyring's, who is also a cousin of my wife's. I believe Henry to be of great integrity and thoroughly honorable. It is my personal opinion that even if no direct slight to Henry Moray's work was ever intended by Carl Eyring, these misunderstandings did him great damage and have been carried on by Henry Eyring - first in his denial that he ever saw the device (Figure 28), and second in his statements that no energy was ever produced. In 1963, when I was informed that Henry Eyring was making these kinds of statements I determined that the only way of ending this misunderstanding would be to have a conference with Henry Eyring. Gene Vickers and I went to see Dr. Eyring and confronted him with the fact that he had seen the lights light and the motors run. At that time he did "admit privately" that he had seen the device in operation. But publicly he would not say that he had seen it (Figure 29). The following letter dated January 14, 1974, was written by Gene Vickers as his account of that day. In February, 1965, Dr. Moray and I wrote a memo for record which carries my memory of the event. Any minor discrepancies that may appear are of little importance; the general information is all that I feel is important (Figure 30).

Also in 1963 in conjunction with his own efforts to develop Radiant Energy, Dr. Robert Craig of Norburg Manufacturing and a friend of Henry Moray from the REA days, wrote Henry Eyring at the University of Utah. In his reply (Figure 31) Henry Eyring denied having ever seen the device. The bottom of the letter contains a handwritten statement by Dr. Moray. Again in May, Henry Eyring told Ed Hermann essentially the same thing.

There are many things in science that work but it has not been proven why

University of Utah
Salt Lake City

Graduate School
207 John R. Park Building

Office of the Dean

May 7, 1963

Mr. Edward W. Hermann
McGraw-Hill Publishing Company, Inc.
500 Fifth Avenue
New York 36, N.Y.

Dear Mr. Hermann:

Since Mr. Moray's equipment was not made available to me for inspection there is really nothing useful that I can say about his work.

Yours sincerely,

Henry Eyring

Henry Eyring
Dean

HE:bb

Fig. 28

Letter from Dr. Henry Eyring to E.W. Hermann

Advanced Concepts, Inc.

Phone (801) 532-5851

P. O. Box 74
Salt Lake City, Utah 84110

January 14, 1974

Mr. John Moray
Research Institute
2505 S. 4th E.
Salt Lake City, Utah 84115

Dear John:

At 10:00 a.m. on February 28, 1963, you and I had an appointment with Dr. Henry Eyring in his office at the University of Utah.

The purpose of our appointment was to discuss Radiant Energy; not only to discuss it, but to obtain Dr. Eyring's comments on his witnessing the demonstration of Radiant Energy. I informed Dr. Eyring of the purpose of our visit, and he agreed to discuss it with us.

To the direct question "Have you seen Radiant Energy demontrated?" he answered, "Yes," that he was present at a demonstration presented by Dr. Henry Moray of what Dr. Moray called Radiant Energy. I then asked him if energy was produced, and he again replied, "Yes," he saw evidence of energy being produced, that he saw lights light up.

I then asked him if he would be willing to make a statement to that effect. He replied, "No." I asked him why. He replied that he had seen evidence of energy produced. He was allowed to examine the device, all except one part that Dr. Moray held in his hand. Because he was not allowed to examine the small item held in Dr. Moray's hand, he would have to say he did not see Radiant Energy in operation because he was not allowed to examine all parts of the device producing the energy, that as a scientist he would have to see everything or he could not attest to the unit working.

I then asked him what difference it made if he didn't see the small part that was capable of being hidden in Dr. Moray's hand, that if the device worked certainly that small part couldn't affect his visual observation of the device producing energy. He replied that as a research scientist he had to know and understand all the working parts or he could not make a statement using his name that he had seen the device in operation. He felt his standing in the scientific community was such that it would be improper to say he had seen the device when he didn't see and understand the operation of all parts.

I then asked him if anyone asked him about it, would he be able to say he had seen it work. He said he would have to say no again for the reason he had given prior, that he had not seen the little part held in Dr. Moray's hand.

Fig. 29

Letter from Gene Vickers

Page 2
January 14, 1974

He told us that he was at a loss to understand private research such as Dr. Moray conducted. The reason was that he had worked in public research, and everything they worked on was made public. Nothing was held back. He felt that each man had the right to pursue whatever research he wanted to, but he could not understand a person pursuing individual research, developing something of such a magnitude as Radiant Energy could be and then withholding the information fram the public.

John, I have placed in quotation marks as near as I can recall Dr. Eyring's exact words. Other statements are recollections of his comments but are not recollections from notes of such extent that I can say these are direct quotes. However, the words I have used convey the meaning of what he was saying.

I hope this is helpful to you in some way.

Sincerely,

Gene

Gene E. Vickers
President

GEV:ilm

Letter from Gene Vickers *(continued)*

they work. The rejection of Moray's work simply because no one could understand why it worked seems almost unbelieveable, to say the least. We still do not understand just why aspirin works, e.g., but that certainly does not prevent its widespread use.

Even more obvious damage was done by Henry Eyring in discussing Henry Moray's work with Commissioner C. E. Larson of the Atomic Energy Commission (Figure 32). William Kerber of Washington, D. C. had asked Larson to look into Moray's work, as a possible alternative to atomic energy. Commissioner Larson stated that he had talked to Henry Eyring and that Henry Eyring was convinced there was "no evidence that any energy has been harnassed by Moray." But if the usage of power is not a sign of energy being harnassed and provided, he must not understand the term "energy." Commissioner Larson also stated he thought that Henry Moray had not permitted examination of his apparatus. On the contrary, as I have argued before, I believe Fletcher and Eyring and several others had received complete descriptions both in their attempts to obtain scientific information on the device and as consultants for Dr. Moray.

Comparing Henry Eyring's statement with that made by Harvey Fletcher, a contradiction appears. I find it very interesting that Eyring said there is no demonstrable evidence of energy being harnassed, but Fletcher said that the result being obtained was too large to be explained by the only theory he was able to find to account for the source of energy, admitting that there was more wattage than they could explain. On the one hand, Henry Eyring is saying that there was no evidence of energy; but on the other hand, Fletcher is saying there is too much energy to come from known sources. In other words, because they didn't know the sources, they discounted any evidence they saw. This is what Henry Eyring explained to me. Because he didn't understand it he would in no way admit that he saw it.

A scientist does not have to say that he knows how something works in order to say that he has seen experimental evidence that it operates. When we examine a research project we can state we have seen a reaction take place even though we do not understand why. If we say that there was no evidence of energy being harnessed, how do we explain the energy? Where could it have possibly come from? Henry and Carl knew it didn't come from any of the components of the device because they examined all the components of the device. Henry Moray's note at the bottom of Henry Eyring's letter of April 25, 1963, says Henry Eyring was given an opportunity to examine the entire device. He was shown how the solid state device, the detector, worked. Although he was not allowed to dismantle the RE detector, this was because dismantling would have destroyed it. However, they were allowed to test for energy coming out of that detector, and they were shown a substitute device and allow to dismantle this. Henry Moray could easily reproduce the radio detector. But to reproduce the RE detector was impossible, forcing him to use some discretion in dismantling this two ounce bit of material.

I believe Carl and Henry Eyring checked every single part of the Radiant Energy Device and determined that none of the components were active. Consequently, the important question for the scientists viewing the demonstration

Salt Lake City, Utah
February 19, 1965

Regarding Gene Vickers' and John Moray's visit with Dr. Henry Eyring in the Doctor's office at the University of Utah in 1963, I can only give an account of this visit from memory after hearing what took place recited to me first by Mr. Vickers (whom I suggest you contact for complete details) and on more than one occasion by John Moray.

Gene (Mr. Vickers) and John called on Dr. Henry Eyring by appointment and John asked the Doctor if he would give them an account of the Doctor's visit at the Moray laboratory and his witnessing the Moray Radiant Energy device in operation under an experimental demonstration.

The Doctor replied, "I never saw anything." John replied, "You never saw anything! What do you mean? You saw the device in operation, saw it heat a flat iron, a glow heater, run a special high frequency motor and light a bank of standard electric light globes." The doctor said "Oh, I saw some lights light but that did not mean anything." John replied, "You saw the device in full operation and made several tests of disconnecting wires, etc. and in the end the device was opened and dismantled part by part for you and you were permitted to examine all the parts." Dr. Eyring said "Yes, but I did not see anything, and I may be a Prof. Anthony and a Dr. Mitchel when I say it but I did not see anything (this expression has been used to describe an adverse attitude around certain Utah people for many years) and my position is still one of

Fig. 30

Memo, Feb. 19, 1965

I did not see anything and what I did see I did not understand."

As I said above, I would suggest you get a statement from Mr. Vickers for first-hand information as I do not wish to bother John in this matter at this time.

I also suggest you read a letter I wrote a Mr. Herman after he had received a letter from Dr. Henry Eyring. It must be remembered Dr. Carl Eyring, after he had spent a day with me and his personal friend, Attorney Robert Judd, and had been very impressed and interested in R.E. to the point of trying to assist in wording a theory, he turned to the induction theory and in several ways made such a spectacle of himself over R.E. that the big bosses of the B.Y.U., who had been considering him for President of the B.Y.U., put another in in his stead.

Dr. Fletcher is also a relation of Doctors Carl Eyring and Henry Eyring, and Dr. Fletcher has made a number of ridiculous and contradictory statements in writing regarding R.E. All this can be proven by documentary evidence and many of these facts have been published in print over a long period of years with no effort on the part of any of these Doctors to refute my statements and the letters of this friend in the matter.

Why is it these gentlemen of high repute, when they cannot refute R.E. from any standpoint, resort to an unfounded, an untrue personal scandal over a period of several years?

John E. Moray

T. Henry Moray

T. Henry Moray

Memo, Feb. 19, 1965 *(continued)*

UNIVERSITY OF UTAH
SALT LAKE CITY

GRADUATE SCHOOL
207 JOHN R. PARK BUILDING

Office of the Dean

April 25, 1963

Mr. Robert B. Craig
Nordberg Manufacturing Company
607 Barr Building
910 - 17th St., N.W.
Washington 6, D. C.

Dear Mr. Craig:

With respect to your inquiry regarding Henry T. Moray, I can only say that since Mr. Moray did not make the instrument available for inspection I have no way of knowing anything about its effectiveness.

Yours sincerely,

Henry Eyring

Henry Eyring
Dean

HE:bb

This statement is 100% untrue as Dr Henry Eyring at his own personal request was permitted to be present at an experimental test The Dr saw the device in full operation; lighting a bank of 100 and 150 Watt lamps and heating electric glow heater, flat irons and operate a 400 cycle electric motor. He also saw the device torn down part by part and examined each part

Fig. 31

Dr. Henry Eyring's Letter to R.B. Craig

UNITED STATES
ATOMIC ENERGY COMMISSION
WASHINGTON, D.C. 20545

JUL 17 1970

Mr. William Kerber
The Westchester
4000 Cathedral Avenue, N. W.
Washington, D. C. 20016

Dear Mr. Kerber:

I have just talked to Dr. Henry Eyring of Salt Lake City (who just returned to town after a long trip) concerning Mr. Moray's project to obtain energy from "radiation." I might explain that Dr. Eyring is a distinguished scientist and is one of very few scientists in the country who has been awarded the Presidential Medal of Honor.

He has seen Mr. Moray's demonstration and is ~~not~~ convinced that there is no demonstrable evidence that any energy has been harnessed. Since Mr. Moray has not permitted examination of the apparatus, there is no way that one can rationally deal with the problem. If there is any merit to the claim, there would be no difficulty in obtaining support from any one of hundreds of concerns who are continually seeking new technological developments.

I am returning the material. I appreciate the opportunity of looking into this interesting matter.

Sincerely,

Clarence

C. E. Larson
Commissioner

Enclosures

Fig. 32

Letter from AEC to William Kerber

became: If the energy was not harnessed from the universe, how could they explain that the lights lit and the motor ran? And no one, not even Fletcher and Eyring could really answer that question. Thus, in both Eyring's statements, first in his letter of May 7, 1963, to Ed Hermann ("Moray's equipment was not made available to me for inspection"), and then in his letter to Dr. Craig, April 25 1963, he said the same thing: essentially, that Moray did not make the instrument available for inspection. Gene Vickers paraphrases him by saying, "He didn't see the small part that was capable of being hidden in Henry Moray's hand." But the part was **not** hidden; he **was** given an opportunity to examine it, he knew it was not the source of power. Henry Eyring is not in the same situation as Carl Eyring or Harvey Fletcher — Carl Eyring and Harvey Fletcher were consultants.

Having discussed this once with Henry Eyring, I do not feel myself under any obligation to discuss it with him again.

One can easily see from the evidence presented that it was difficult for Henry Moray to communicate effectively with Dr. Henry Eyring. Because he did not understand the Radiant Energy device, Eyring seemed convinced that he could not **afford** to say that he saw it operate.

In 1961, Henry Moray and I traveled to Washington, D.C. at the invitation of Congressman David King to visit NASA. By this time, Robert Craig and Moray had become very friendly. Dr. Craig met us and agreed to represent Moray to negotiate an attempt to get NASA to develop Radiant Energy. Robert Craig, having been with the R.E.A., was thoroughly convinced of the great and general value of Henry Moray's work.

On July 24, 1961, in the morning, Dr. Craig got Dr. Moray out of bed to inform him NASA had just called to say that they would be willing to go forward on the five million dollar program we had outlined, with the stipulation that after the first $2.5 million had been spent, the project could be re-evaluated. However, in a letter received from NASA, dated 21 July, the assistant director, Mr. Colon, stated: "We appreciate your offer to have some of our scientists visit your laboratory in Salt Lake City with a complete review of the work which you have done on Radiant Energy. However, we regret that we must advise you that after careful study of your proposal by scientists both in this headquarters and in our research center most active in the science area involved, it is our considered opinion that the probability of your Radiant Energy concept resulting in a practical, useable device for aeronautics or space flight applications is not sufficient to justify support by this agency." Dr. Craig, in further investigating this rather complete NASA reversal, was told by friends that the Appropriations Committee - i.e., Utah Senator Wallace Bennett - had killed the project. Consequently, I went rather irately to a friend of mine, Lyle Holmgren, and requested Lyle to ask Senator Bennett for an explanation.

Senator Bennett's letter is enclosed (Figure 33). In it the Senator takes the attitude that Dr. Moray asked for something for nothing and that it only involved a minimum of $50,000. And that Moray refused to reveal the whereabouts of his "box." This is particularly interesting as the story relates to nothing I am aware of, and I was present during all negotiations with NASA. In

WALLACE F. BENNETT
UTAH

L. RALPH MECHAM
ADMINISTRATIVE ASSISTANT

COMMITTEES:
FINANCE
BANKING AND CURRENCY

JOINT COMMITTEES:
ATOMIC ENERGY
DEFENSE PRODUCTION

United States Senate
WASHINGTON, D.C.

P. O. Box 1320
Salt Lake City, Utah
October 30, 1961

Mr. Lyle E. Holmgren
Holmgren Building & Trust
29 West Center Street
Logan, Utah

Dear Mr. Holmgren:

When I received your letter of October 26, I asked my staff in Washington to check with the National Aeronautics and Space Administration to find out what, if anything, had happened as a result of Mr. Moray's earlier presentation. The reply I get is very discouraging.

Apparently Mr. Moray has a big box somewhere which he says will give up radiation, and that by some means which he won't divulge he can direct this energy where he wants it to go. He has asked NASA to give him a minimum of $50,000 to use in any way he pleases, but flatly and steadfastly refuses to reveal the whereabouts of the box or to give the scientists in government any information about his process. I am sure if you as a businessman were approached by someone with a similar proposition, you would be equally skeptical. I am equally sure there is no "conspiracy," by which I presume that Mr. Moray feels that if he reveals any of the basic scientific theories or data behind his claimed discovery this will be "stolen" from him and therefore he is asking the government to buy a pig in a poke.

If Mr. Moray had the true scientific approach, he would not be afraid to reveal his program to reputable men of science, men who are also men of honor. He would probably also have moved to get a patent on his program, and of course this requires revelation of its basic aspects.

I am afraid that as long as he insists on keeping this as a tight secret, asking people to trust him in the face of the fact that he is unwilling to trust them, there is nothing that can be done to bring him and the government together.

Fig. 33

Letter to Lyle Holmgren from Sen. Wallace Bennett

Mr. Lyle E. Holmgren
Page 2
October 30, 1961

According to the information that has come to me from Washington, he has been trying to sell his proposition to the government for many years and has been repeatedly to many agencies, and in no case has he been willing to take anyone into his confidence.

I am sorry I can't give you a more favorable reply. Thank you for taking the time to write.

Sincerely,

Wallace F. Bennett

WFB:j

Fig. 33

Letter to Lyle Holmgren from Sen. Wallace Bennett *(continued)* the third paragraph he questions whether Moray has a true scientific approach and implies that Moray should simply openly release his work. My father refused to make specific disclosures without signed and legally binding contracts. If this is carrying an invention as "too tight a secret," then why do patent laws require it.?

Several other competent observers wrote Henry both personal and professional letters. Dr. Frank Stewart Harris, at the time president of Brigham Young University, wrote making reference to his visit with Henry Moray. Years later Dr. Harris seemed not to remember visiting Henry and seeing the device in operation. A number of letters were then exchanged, and after reexamining his files Dr. Harris wrote a very friendly letter.

Dr. Milton Marshall of Brigham Young University remained friendly; he wrote several letters which show his interest in Dr. Moray's work and their close association.

Another doctor of science who was interested in Dr. Moray's work and willing to publicly say so was Dr. O.L. Polly, physicist from Baltimore, Maryland, who wrote in 1936 of his visit to the laboratory (Figure 34).

It would be quite improper for me not to say something positive about Henry's friends. S.E. Bringhurst entered the picture at the end of the Moray Products Company fiasco. Having been one of the stockholders, he became one of the officers after Schade, Farnsworth and Hayes were eliminated. He worked with Henry as the corporation was dissolved, and he was vice president at one time.

As the Research Institute developed, S.E. Bringhurst was the first president of the Research Institute. He was a businessman in Murray, Utah, and very successful. He, like the Moray family, shared some of the burdens. Shots were fired at him within a few weeks of the time my mother and the rest of the family were fired at. S.E. did not have bullet-resistant glass, and the bullet passed by the right side of his head and exited the car at the rear.

All during the R.E.A. days, S.E. Bringhurst gave my father close support against those individuals who were most officious in their administration of the program. Until about 1942, S.E. devoted considerable effort to raising funds for Dr. Moray to finish his research and standardize the Radiant Energy device. S. E. became a mission president for the Church of Jesus Christ of Latter-day Saints in August of 1944, and he resigned as president of the Research Institute at that time. His successor was Chester M. Todd.

Chester Todd, Henry's nephew, has never ceased to be enthusiastic about Radiant Energy. To this day he is one of its greatest active supporters. He is presently the principal of Churchill Junior High School in Salt Lake City and a director of the Cosray Research Institute. Chester lives in the original home his mother and father built on Fifth East Street next to where Henry lived.

Another person that I certainly shall not forget is William Lovesy (Figure 35). Bill Lovesy's niece, Afton Henegar, often spoke of how enthusiastic Bill Lovesy was about Radiant Energy and how positive he was that it was a force with the "power to change the world." Henry often quoted Bill as saying, "I can go so far to interest people in Radiant Energy, and then it seems as though I run into a stone wall." Bill came into the picture at the end of the Moray Products Com-

— Letter from Dr Polly

Baltimore Md.
Nov. 8 1936.

To Whom It may Concern:

On the occasion of a recent visit to Salt Lake City (Sept 1936) I arranged, because of my own interest in various reports I had heard, to visit with Mr T.H. Moray and discuss with him some of his experiments.

I was happy to find his laboratory well equipped with a number of costly electrical measuring instruments, and soon became convinced that he understood their uses.

His radiant energy device, as he calls it, seems to involve no mysterious circuit, altho the electrical valve upon which it depends is novel. The apparatus does not depend on any battery or other source of electricity than the antenna. In other words the power obtained is not derived thru the medium of a secondary power source — no feeble current is built up as in a radio set by vacuum tube amplifiers. However, no violation of the laws of thermodynamics is implied.

practical interest. I saw no evidence that he was perpetrating a hoax and I gladly write this in appreciation for the time he spent with me.

O.L. Polly.

Dr Polly received his degree Ph D in Physics from College in ~~Science~~ Baltimore Md.

Fig. 34

Letter from Dr. O.L. Polly

William Lovesy

Fig. 35

pany era, and he corresponded with Murray O. Hayes, Harvey Fletcher and Carl Eyring in an effort to verify the work that Dr. Moray had done up to 1932. Henry often said that if Bill Lovesy had lived, he was sure he would have made Radiant Energy a success. Bill did not take "no" for an answer. He died of injuries received in an automobile accident when returning from Denver. His passenger, a hitchhiker, was uninjured in the rollover. After the accident, Bill lived several weeks in the hospital but was never able to speak again.

K.K. Steffenson was not only an attorney who stood by Henry Moray for many years, he also became the secretary of the Research Institute and remained an officer of the corporation longer than any other. He was still secretary of Cosray Research Institute at the time I became president. K.K., as we called him, seemed to be ostracized by the business community of Salt Lake, though I never understood why. As an attorney and as a man his integrity was impeccable, a person having the highest intelligence, having succeeded in a number of academic fields and being physically very apt in a number of sports. It is my understanding he was a personal friend of Franklin D. Roosevelt's from Columbia University. His notes and his books are still stored in the Cosray laboratory since they were purchased by my father in an effort to preserve them, with the idea that at some future date they could be published. K.K. had a favorite joke that I've often enjoyed telling: "In Salt Lake City, things are becoming so bad they are burying two people in one grave. On one of the headstones there is a notation, 'Here lies an attorney, and an honest man.' "

Another friend that I remember specifically was Nathaniel Baldwin. Baldwin was the founder of what became Baldwin Radio. When Baldwin Radio was in Salt Lake City, Nathaniel Baldwin had a considerable amount of money. Baldwin wanted to support Radiant Energy. However, Henry felt that Nathaniel was tied up with too many people who were leading a course to disaster. History proved Henry to be correct. Nathaniel Baldwin went to jail under conditions such that even the judge make the remark that he believed Nathaniel Baldwin to be innocent of any intentional wrongdoing, but as president of the corporation, he was responsible for actions of those under him.

After Nathaniel left prison, he used to walk about four miles to come and see my father at the laboratory on Fourth East. I remember many times going to the door and seeing the old man in a ragged coat standing in the cold. After a visit, my father would say, "Nathaniel, I was just sending Eugene on an errand, and he is going right by your place. Let him take you home." Between the two men there was the greatest respect. Dad respected Nathaniel for his honesty and integrity. Nathaniel Baldwin respected my father because Henry never gave up even though circumstances turned against him. Henry remained Nathaniel Baldwin's friend until the day Nathaniel died.

Gabriel Mes

Fig. 36

Many persons have questioned how Henry could have possibly built such an intricate device as we have reported it to be. This was due in no small part to the assistance of Nathaniel Baldwin in making available the facilities of Baldwin Laboratories. Also, William Lovesy brought the facilities of the Great Western Radio, located in Salt Lake City at that time, to Henry. These two laboratories furnished a great amount of material and labor without obligation.

Gabriel Mes (Figure 36), a master machinist and a wizard with anything mechanical, worked in a machine shop on East Second South in Salt Lake City. Many times my father took me with him to see Gabriel Mes, and I remember the big lathes and the wonderful equipment that he had. I remember going with Henry and Gabriel when they used to recharge the magnets on the magnetizer that Gabriel Mes had at the facilities of Great Western Radio. I remember the stories Gabriel used to tell about his work as a machinist and about some of the wonderful work he had done for my father and others. He was proud of his work. He was able to machine two pieces of steel so that without oil they would stick together from the vacuum created by the finely machined surfaces.

Alfred Burrell was more than an acquaintance and a relative, he was a good friend. Alfred was a watchmaker and jeweler, and he did much of the fine work my father was unable to do. The adjusting of the cat whiskers and the fine soldering in building detectors were done by Alfred Burrell. Alfred is now in his eighties and to this day he speaks fondly of my father and the work he did for him, and of how the Radiant Energy Device amazed him.

I remember George Danly, the founder of the Wyoming Lebarge Oil Company. He gave my father some of the oil stock, and it proved to be one of the best assets my father had in his later years. I always knew Danly as "the gum man" because, besides representing the oil company at the time, he also represented one of the chewing gum companies in the Salt Lake City area. Another gum salesman was V.M. Stock, the representative for Beechnut gum. Stock desperately tried to arrange finances for Henry in the early days, just before World War II. However, he was transferred to California by Beechnut, and wartime travel restrictions prevented us from further social contact.

There are several other persons I must mention, and I am sure that there were many that I was not aware of as a boy. Walter Scoville, another oil man, made millions of dollars at various times, but he also lost millions of dollars at various other times. Walter tried very hard to make a go of the business for Dad. Another man who assisted Dad was Harry Harris, who I believe died of leukemia in Tooele, Utah. Another person who befriended Henry was a man named Nalder from the San Francisco area. Nalder was an undertaker and very successful.

I must mention C. R. Benzil from Greeley, Colorado, and also Ralph Noffsinger and his nephew, Harold Noffsinger, from Salt Lake City, who did many of the patent application drawings during the Moray Products Company days. Lou Gardner, Clyde Gardner, E.G. Jensen, Harold Jensen, Grant Stringham - and others I have been unable to remember because of my young age - aided my father. All these men in one way or another supported Henry Moray in his work.

Without Gabriel Mes's contacts in England and Germany, Dr. Moray would have never been able to have his "tubes" made. The reader must realize I use the words "tubes" loosely. Basically my father's "tubes" were solid state devices.

As the years have passed, many well-meaning persons have come, assisted my father's work, and gone. And in this chapter let me acknowledge our appreciation to all those who have assisted or aided in this work. And particularly let me say I deeply appreciate those who in the last few years have honestly tried without guile to further the work.

Finally, I am also impressed that many individuals have come deliberately trying to hinder the work; or if they were not trying to hinder, they fall strictly into the class of "frienemies". One of Henry's "frienemies" once said, "Leave Henry alone; when we need his Radiant Energy Device, he'll be humbled sufficiently so that we will be able to get it."

I don't know how far a man is expected to be humbled; however, I do know that because of such attitudes, humankind has come very close to losing T. Henry Moray's Radiant Energy altogether.

I would like to tie a few remaining loose ends together and show what to me are some of our best letters. I gathered sworn statements over the last few years, because I could see that Henry Moray's work was completed as far as he himself was going to be able to take it. To once again reaffirm what had actually been done, I went to a number of individuals who had seen the device operated.

The first person I approached was Sam Bringhurst (Figures 37 and 38). I showed him a picture and had him swear that this was a picture that he had seen (Figure 39), and on the back of the picture (Figure 37) you will see that he swore in front of a notary, J. B. Bell, that this was one of the tests he saw. S.E., as my father fondly called Bringhurst remained until he died a very close friend of my father's. The two men died not more than six months apart.

The second photograph is an enlargement of a photograph of my father in the basement of his home prior to the time the basement was remodeled for a small laboratory. The reader will notice the crude covering over the windows. This photograph is sworn to be my cousin, Chester M. Todd, principal at a local junior high school and Colonel in the U.S. Army Reserve. The photograph and Chester's sworn statement are included. (Figures 40, 41 and 42.)

The next statement I include is a sworn statement by A.B. Jensen, taken before K.K. Steffenson, in which Jensen gave an account of the work that was done in the presence of, supposedly, a representative of President Hoover. The induction story was still extant, although the battery thesis had pretty well been laid to rest. After all, if one had batteries that were able to put out the kind of power demonstrated over the period of time that the device had operated, the batteries alone would have been a commercial bonanza.

In 1938 Henry Moray wanted to make one more isolated test and decided on what now is known as Dugway, west of Salt Lake City, near Johnson's Pass. Note that even though this happened 25 years before the certificate was made, A.B. Jensen had a vivid recollection of the test (Figure 43).

The next affidavit is one given by David I. Gardner at Midvale, Utah, 14 September 1963, in which David Gardner reiterates what he saw in 1936. He also mentions a few of the other Moray inventions that he was familiar with and had seen (Figure 44).

The next affidavit is a little older. This is an affidavit that A.B. Jensen swore before K.K. Steffenson, December 1956 (Figure 45).

Finally, I have included a statement from Dr. Robert B. Craig. Dr. Craig points out in 1963 that when he was a member of the REA he had personal knowledge of the Moray experiment with the REA, and the engineer, Crim, also

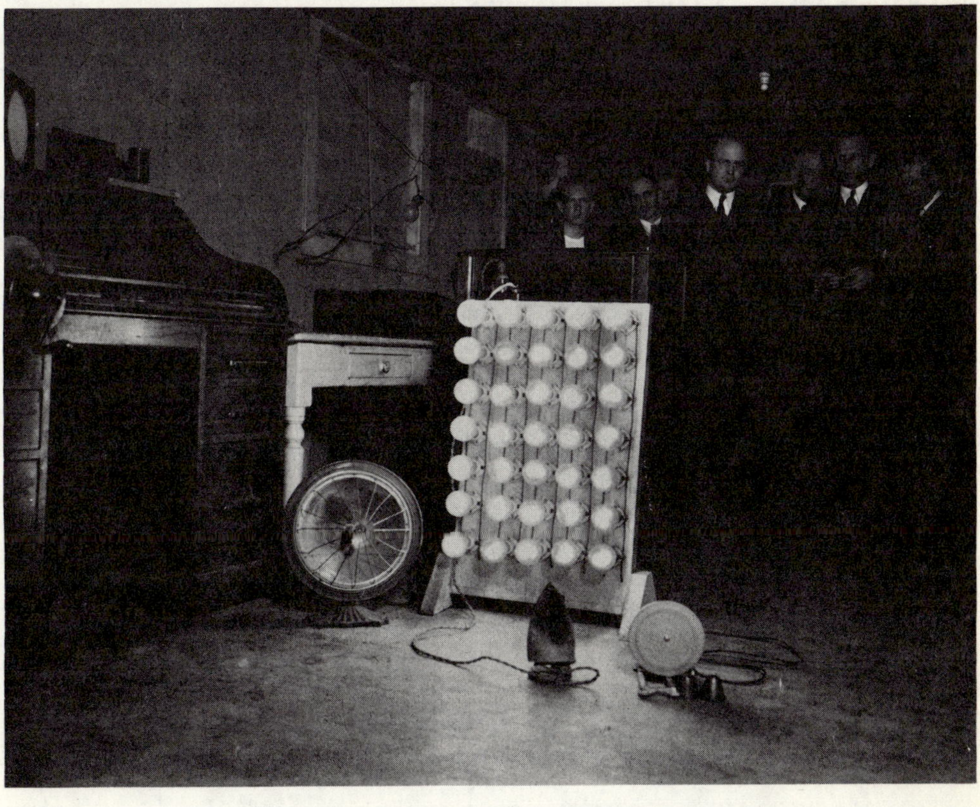

S.E. Bringhurst photograph of Dr. Moray's Lab

Fig. 37

This is one of the tests I saw.

Samuel E. Bringhurst

State of Utah
County of Salt Lake } ss

Subscribed and sworn to before me, a Notary Public, this 19th day of March, 1971.

Jacque B. Bell

Com. Ex. 11-1-71
Residing: Salt Lake City, Utah

Fig. 38

S.E. Bringhurst statement on photo *(See Fig. 37)*

850 Three Fountains Drive
Unit No. 198
Murray, Utah 84107
March 19, 1971

TO WHOM IT MAY CONCERN:

During the later part of July, 1938, I witnessed an experimental test made by Dr. T. Henry Moray at his laboratory which was then at 2484 South 5th East, Salt Lake City, Utah. This was only one of the many experimental tests I had witnessed with various groups of men present.

At this experiment there was a bank of lights consisting of 35 standard electric light globes, 20 of which were 150 watt globes and 15 of which were 100 watt globes. There was also heated to full capacity a 600 watt glow heater, a standard 575 watt flat iron and a special motor which had been designed by Dr. Moray. A picture of this experiment was made in my presence by a commercial photographer, with some of those present being in the picture. I would like to call attention to the electric globe which did not light which was in series with the device when it was in full operation. It will be noted in this picture, which is true of all that have been taken of this device in operation, the whiteness of the light which came from these standard commercial light globes.

I have witnesses many of these tests which have been more or less alike. These tests were made to prove various improvements made in the device from time to time.

I have also read many of the letters of those who have also witnessed various experimental tests of this device in operation.

Samuel E. Bringhurst
Samuel E. Bringhurst

STATE OF UTAH)
County of Salt Lake) ss.

Subscribed and sworn to before me this 19th day of March, 1971.

Jacque B. Bell
NOTARY PUBLIC
Residing at: Salt Lake City, Ut.

Com. Ex: 11-1-71

Fig. 39

S.E. Bringhurst sworn statement before J.B. Bell

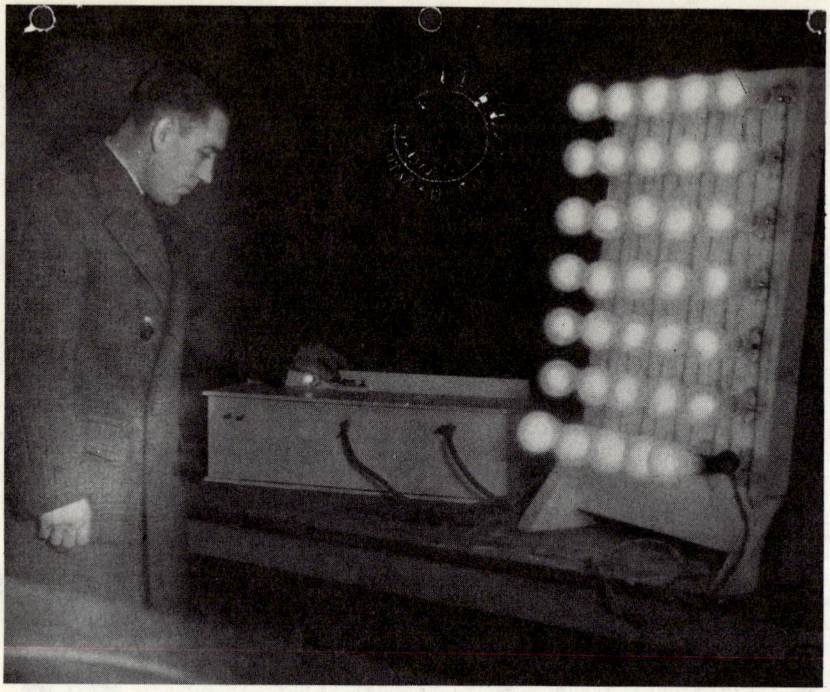

C. Todd lab photo

Fig. 40

This is the test I saw in Feb 1937.

Charles M. Todd

State of Utah ;
County of Salt Lake ; ss

Subscribed and sworn to before me, a Notary Public, this 19th day of March, 1971.

Paul H Peterson

Com. Ex. 7-17-1975.

Residing: Salt Lake City, Utah

Fig. 41

C. Todd statement on photo

2508 South 5th East
Salt Lake City, Utah
March 19, 1971

TO WHOM IT MAY CONCERN:

On a Thursday evening in mid-February 1937, while I was a student teacher at South High School, I called my uncle T. H. Moray to allow me to bring my directing teacher, Mr. Vivian Decker, and my dentist, Dr. Fred K. Bischoff, to witness a demonstration of his radiant energy device. Accordingly, we three did have this opportunity at 2484 South 5th East, Salt Lake City, Utah. The device was housed in a wooden box something like 12" by 18", with an antenna and a ground going into it. Wires leading out of the box led to a bank of some forty 100 watt light globes and to an electric iron. My uncle touched a switch on the top of the box with a hand electro-static plate and the globes all lighted brilliantly. We all noted that the bulbs burned cold except each had a hot spot about the size of a dime on the top slightly off center. I also recall that I could turn the lights on and off by approaching and retreating to and from the device, either with my whole body or my hand. If my memory is clear, the machine had to be tuned with a dial to be placed in this condition.

I clearly recall leaving with the firm conviction that here was a source of free power and a burglar alarm.

Chester M. Todd

STATE OF UTAH) ss.
County of Salt Lake)

Subscribed and sworn to before me this 19th day of March, 1971.

NOTARY PUBLIC
Residing at:

Fig. 42

C. Todd statement sworn before Notary Public

Salt Lake City 6, Utah
December 28, 1962

Dr. T. Henry Moray
Salt Lake City, Utah

You recently asked if I would recall a test made during 1938 of the Moray Radiant Energy Equipment. To the best of my recollection, the test was as follows:

I accompanied a group of people consisting of Mr. J. C. Jensen, Mr. Greely Snyder, Dr. Moray, and myself (I recall one other person also was with us but will not give the name, as I am not sure whether I am correct). We drove out past St. Johns in Tooele County, Utah, and then 25 to 30 miles southwest of Johnson's Pass. At that time there was no installation at Dugway, no telephone or electric wires in the vicinity - merely desert land.

Dr. Moray had packed his equipment and tools necessary to erect it. We selected a spot 50 to 75 feet from our automobile. We assisted in driving a steel rod about 6 feet into the ground. With a blow torch, wires were soldered to the steel rod. We erected two tent poles and fastened an antenae made of #10 stranded copper wire.

The wire from the antenae was fastened to the box containing the Moray Radiant Energy Equipment and the ground wire that had been soldered to the steel rod was also fastened.

Dr. Moray had brought a board with about 35 150-watt lamps mounted thereon, a high frequency motor, a 1000-watt taylor's iron, and I believe a heater or fan.

After connecting the "ground" and "antenae" to the box, Dr. Moray "plugged in" the lamps, iron, motor, etc. The energy from the box lighted the lamps, heated the iron, ran the motor, etc.

We all examined the equipment and determined there were no batteries; there were no hidden connections to electric outlets, as we were miles from any power lines; Mr. J. C. Jensen drove the automobile and selected the location for the demonstration, thus there could have been no prior preparation at the test site by Dr. Moray.

Although the demonstration took place nearly 25 years ago, I still vividly recall the reaction of all who witnessed it. It was our firm conviction that the Radiant Energy Equipment was truly as represented by Dr. Moray.

Very truly yours,

A B Jensen

A. B. Jensen

STATE OF UTAH)
COUNTY OF SALT LAKE)

A. B. Jensen, being first duly sworn, deposes and says: That he has read the foregoing statement and acknowledges that he wrote and signed the same as above set up.

A B Jensen

Subscribed and sworn to before me this 31 day of January, 1963.

K K Steffensen

K. K. Steffensen--Notary Public--
residing at Salt Lake City, Utah

Fig. 43

A.B. Jenson sworn statement

AFFIDAVIT

STATE OF UTAH)
:ss.
County of Salt Lake)

DAVID I. GARDNER, being first duly sworn, deposes and says: That I am a licensed Civil Engineer in and for the State of Utah, License No. 427, and hold a degree of B.S. in Civil Engineering from the Utah State University; I have been a practicing Civil Engineer for the past thirty-seven (37) years, District Engineer for Salt Lake County Water Conservancy District, District Engineer for Granger-Hunter Improvement District, and various other positions.

I am also an inventor with letters patent on a ditch lining device and a patent pending on a service connection on a water main; I also designed and built a dredge boat which is successfully operating.

That sometime about the year 1936 I was invited to visit the laboratories of Dr. T. Henry Moray in Salt Lake City, Utah. Upon my arrival, there were many other people present to see this experimental demonstration of Dr. Moray's. During the course of the evening Dr. Moray demonstrated what he called "radiant energy" consisting of a battery of approximately forty 200 watt globes, small electric motors, flat irons, and electric fans connected to a board wired conventionally with outlets for each globe and electrical appliance. The demonstration consisted of stringing a small piece of thin copper wire of very small diameter across the room. Connected on each end of the wire were two porcelain insulators small enough to hold conveniently in one's hand in order to pull the wire tight. I was given one end of the insulator and someone at the other end of the room was given the other. I was advised by Dr. Moray to pull the wire tight which I did. I asked him what would happen if I touched the wire. He said nothing would happen except it would turn the lights off. I asked him if it would shock me and he answered, "No". We then pulled the wire tight and Dr. Moray

DAHL AND SAGERS
ATTORNEYS AT LAW
17 EAST CENTER STREET
MIDVALE, UTAH

Fig. 44

David Gardner Affidavit

turned on the switch or did some other operation which I do not at present recall, but the lights came on in full force and the electrical appliances functioned beautifully. Being of an inquisitive mind, I wondered if he was telling the truth so I carefully let my hand touch the small copper wire, and the lights went off as Dr. Moray said they would. Dr. Moray said that someone was touching the wire; I took my hand off the wire and the battery came on and the appliances again functioned. How it worked I cannot say, but that it did work I am sure. There was, of course, much discussion of the phenoma that we had witnessed among the crowd.

The gentleman who had invited me to the demonstration mentioned a contrivance that Dr. Moray had also invented with which he could pick up sounds without them being broadcast from a radio sending station. He said that he had heard a demonstration of this device. I asked if I might hear it also; he called Dr. Moray over and asked if I might see a demonstration of the "sound pickup device" and I was granted the privilege.

There was an old radio box of conventional design with two headphones plugged into the box. Dr. Moray tuned the device by turning the knob in front. He handed me one set of headphones and one set to my nephew Mr. Clark Gardner; he then arranged for three people to walk out of the house into the street and carry on a conversation to which we were tuned. One of the people chosen to walk from the house was my brother Archibald Gardner; the others were Louis Gardner who had invited me to see the demonstration, and Mr. Albert Dimond who I knew well and whose voice I could determine. I was sure I was listening to these three men. I remember distinctly it was raining at the time and I could hear the patter of the rain on the pavement as they walked. I remember them saying they had better hurry back into the house lest they get wet.

In the course of their return, I wondered if I might hear other people talking and despite the advice of Dr. Moray not to fool around with the tuning

Fig. 44

David Gardner Affidavit *(continued)*

device, I moved the knob slowly and heard very distinctly the sounds of a railroad station: the whistle of the train, and a porter or the station master at the station saying, "All aboard". The nearest station to the receiving set would have been a minimum of five (5) miles. I knew not what station I had tuned in on or what railroad, but the people were talking the English language.

I know that the people walking on the street had no transmitting device or any other instrument with them. They would not believe me when I told them what I had heard, but they said I was repeating their conversation.

David I. Gardner

David I. Gardner

Subscribed and sworn to before me this 14th day of September, 1963.

Joanne B. Rose
NOTARY PUBLIC

My commission expires: 9-20-64

Residing at: West Jordan, Utah

Fig. 44

David Gardner Affidavit *(continued)*

December 22, 1956

To Whom It May Concern:

In the spring of 1938, I was invited by the Securities Commissioner of the State of Utah to accompany Mr. Gull to the laboratory of Dr. Henry Moray and inspect the Moray radiant energy equipment. My father, Mr. J. C. Jensen, an engineer, was also present. He has seen many tests made of the Moray equipment. This invitation was given me because I am an engineer and battery expert.

After seeing the Moray equipment light 30 one-hundred-fifty watt standard light globes, heat a 1000 watt sodering iron and operate a special high frequency motor for some time over an hour, I placed my measuring instruments on the antenna and ground connection. They were electrically dead. They were checked in detail for hidden wires, there were none. The box in which the equipment was housed was opened by Commissioner Gull and myself. Every part of the equipment was taken out of the box, weighed and tested. Tests were made with sensitive volt and amp meters to determine if the box could contain batteries or if any of the equipment had any sort of battery action in them. They did not.

There were no hidden wires and nothing from which any battery current or any other power could have been obtained other than as claimed by the inventor.

The commissioner then requested permission to bore holes in the box in which the equipment had been housed in order to satisfy himself that everything had been taken out and there were no hidden compartments. This was done to the complete satisfaction of all present.

I have been asked for my comments concerning tests and demonstrations of Dr. Moray's radiant energy equipment, and the above is a review of the test and demonstration which I witnessed. I endorse this equipment as represented by Dr. Moray.

A. B. Jensen
425 Felt Building
Salt Lake City, Utah

STATE OF UTAH)
COUNTY OF SALT LAKE)

A. B. Jensen, being first duly sworn, deposes and says: That he has read the foregoing statement and acknowledges that he wrote and signed the same as above set up.

Subscribed and sworn to before me this 3rd day of January, 1963.

K.K.Steffensen--Notary Public--
residing at Salt Lake City, Utah

Fig. 45

A.B. Jensen sworn statement

had personal knowledge of what had taken place (Figure 46).

The Craig letter is particularly interesting; in the next chapter the reader will discover that Henry Moray's differences with the REA were very great. And yet, in spite of these differences, Dr. Craig - who had not participated in some of the covert activities of the REA - made his peace with Henry and proved to be one of his greatest supporters.

I've also added two other interesting letters, not necessarily in chronological order. The first is a letter dated June 29, 1932, written W.H. Lovesy (Figure 47). Mr. Lovesy was unafraid to tackle the so called "experts," and he attempted to pin down both Fletcher and Eyring in order to establish once and for all their position and to get financial backing for the project. Both gentlemen had evidently had second thoughts about the Moray device by this time; and the agreement between Mr. Lovesy and these two gentlemen was that if Dr. Murray O. Hayes were given a complete disclosure, then they would back it. Hayes was given this complete disclosure. I quote from the letter he gave Mr. Lovesy, October 24, 1929, in which he states: "You are already aware that I have seen many demonstrations of what this mechanism will do and that I have seen the parts of which it is built. Recently, Mr. Moray had shown to me the wiring diagrams and assembly, and I am free to say that I can find no inconsistencies in it nor anything which does not appear to be logical and sound. While the hookup appears to be very complicated when looking at the machine, it is in reality very simple. In essence, it is based upon recognized laws of electricity when all is explained. There are many features which appear to be incidental, but they are in reality of basic importance.

"He has also shown me and explained the detector which he uses. In this he has applied a fundamental principle of electrical circuits, which I believe would not be noticed unless pointed out by him. This element of his device also, as above mentioned in respect to the circuit, has numerous features which appear incidental but are the heart of the matter and of first importance."

So the idea that Dr. Moray made no complete disclosure and that no one was taken into his confidence is not valid, and this story, like the G. E. sellout story, was generated by his enemies simply to stop his progress.

Henry Moray made one other attempt to convince the scientific community that his device was genuine. He had already completely disclosed it, not only to his patent attorney, but to Hayes, Fletcher, and Carl Eyring. He had offered to do the same thing for Millikan; however, Millikan had refused, saying he would take Eyring's and Fletcher's words. but now he agreed to let Dr. Knudson, of Stanford University, examine the device.

A lawyer arranged for Dr. Knudson to observe some tests and to perform other tests on the RE Device to establish that the device was operating properly. I know that my father became very upset because Dr. Knudson repeatedly threw the switch in and out, resulting in an inductive surge that burned out the device.

Dr. Knudson left without fulfilling his promise, given to Dr. Moray and Mr. Richards, to write a statement admitting what he had seen. Consequently, when Dr. Knudson did not do as he had promised, Dr. Moray wrote the enclosed letter objecting to what had happened (Figure 48, 49, 50).

NORDBERG MANUFACTURING CO.

NORDBERG MACHINERY

DIESEL ENGINES—SYMONS CRUSHERS AND SCREENS—MINE HOISTS
R. R. TRACK MACHINERY—HYDRAULIC PRESSES
BASIC PROCESSING AND SPECIAL MACHINERY

Cable Address: "NORDBERG MILWAUKEE"
Milwaukee Telephone SHERIDAN 4-2340

REFER TO SUBJECT OR INDIVIDUAL WRITING

TELEPHONE: 296-3640 / 296-3641 / 296-3642
TELETYPEWRITER: WA-239

MILWAUKEE, WIS. U.S.A.

ADDRESS ALL COMMUNICATIONS TO THE COMPANY

WASHINGTON OFFICE:
907 BARR BUILDING
WASHINGTON 6, D. C.

August 22, 1963

Mr. Everett E. Dahl
Dahl and Sagers
17 East Center Street
Midvale, Utah

Dear Mr. Dahl:

With reference to your letter of August 13th inquiring my knowledge of the radiant energy experiments which the REA had with Dr. Moray, I wish to advise you that the reports prepared by several of the REA engineers were put in the Administrator's files.

I think these covers periods for 1938, 1939 and may even enter into 1940. The most important report that I remember is by a Mr. Ben Crim, a senior regional engineer in REA, who went to Salt Lake City to observe certain tests of Dr. Moray at his laboratory. Mr. Crim reported that the tests were entirely satisfactory as far as performance went, but he was not able to explain the phenomena by which the results were produced. There were other reports later to the same effect. I do not have copies of these reports and I am sure that if they still exist in the files of the Administrator of REA, they are in dead storage, but I have a very good hunch that they have either been lost or destroyed. You will remember that REA was transferred to St. Louis in 1941 from Washington and over a carload of files were lost and never recovered and, in fact, so great was the confusion created by this move and the subsequent return to Washington after the war, that much pertinent material concerning the early days of REA and even personnel records have been lost. I am sorry I cannot be of more help to you than this, except to say that I personally observed on several occasions the operation of Dr. Moray's device and like Mr. Crim, I can testify to the results, but not to the how.

Sincerely,

Robert B. Craig
Robert B. Craig

RBC/hg P. S. Mr Crim died several years ago
RBC

This letter refers to tests made by United States Government Scientists and Engineers. Dr. Craig was Assistant Administrator of the R.E.A.

Fig. 46

Letter from Robert B. Craig

Utah Oil Refining Company

Salt Lake City, Utah

W.H. LOVESY
Purchasing & Traffic

June 29, 1932

Mr. T. Henry Moray
2484 South 5thEast
Salt Lake City, Utah

Dear Henry:

I have handed you the original letter of Murray O. Hayes, dated October 24, 1929, in which he advised of his being shown and having explained to him the wiring diagram of the assembly, and also the fundamentals of the detector, and in which he stated plainly that while it would not have been noticed unless pointed out by you, at the same time he had become familiar with all of the detail of the construction of your device for utilizing cosmic energy, and added that you had applied only fundamental principles of electric circuits.

There is attached a second letter in which Mr. Hayes furnished me a detail of his educational record.

The memorandums on the face of these two original documents were made by me at the time I was consulting with Murray O. Hayes in my office and, at the time, he definitely advised me that he had been shown every detail of your invention and added that he could make one of the machines himself from the information that had been given him.

Murray O. Hayes made these similar statements to me several times, and it may be from my determination in having him repeatedly make the declaration that he had been shown every detail of the invention that would give him the impression that I had a doubt about the invention.

Personally, I have never had a doubt, but I thought the real detail secrets of the invention were held only by yourself and after my conference at New York with Harvey Fletcher and Carl Eyring, at the Bell Laboratories,- having agreed with them that I would endeavor to induce you to give the detail of your invention to Murray O. Hayes,- I did, as I have stated, repeatedly ask Murray O. Hayes if he could duplicate your machine, and in every instance his answer was emphatic and to the effect that he could do so.

Yours truly,

W.H.L.

WHL/T

Dr. Murray O. Hayes is a Scientist with a PhD degree as well as a patent Attorney

Fig. 47

Letter from W.H. Lovesy

December 19th, 1931.

Dr. T. H. Moray,
2484 S. 5th E.,
Salt Lake City, Utah.

Dear Dr. Moray:-

This will acknowledge receipt of your letter of December 16th, 1931. I was very glad to hear from you. I have not received a copy of your three booklets to which you refer and appreciate your offer to send same to me.

With reference to Dr. K, you will remember the demonstration that was given him was interrupted by the valve stopping its function and we were hoping that you would be able to start its operation again so that his check could be completed. I would very much appreciate knowing whether you have been able to start the device since that time and whether it is now in operation. Kindly let me know this by return mail. I am expecting to be in Salt Lake soon and will talk to you personally about the matter at that time.

With kindest personal regards and wishing you the Season's Greetings, I am

Yours very truly,

J.D. Richards

Fig. 48

Letter from J.D. Richards

This letter was written to Mr. Richards because he had been instrumental in having this test of the Moray radiant energy device witnessed by Dr. Vern Knudson of Stanford University. I gave this interview to Dr. Knudson and let him witness this experimental test because he had promised that he would, if he were permitted to be present when this experimental test was run, give me a letter stating exactly what he had seen and his reactions tc the test. When Dr. Knudson failed to give me the promised letter, I wrote the following letter to Attorney Richards in Los Angeles, California. He also maintained offices in Salt Lake City under the firm name of Clark, Bowen, and Richards. Original copy of letter follows.

Salt Lake City, Utah
December 22, 1931

Dear Mr. Richards:

I note with interest your reference to Dr. Vern Knudson as Dr. K. and the statement that the "demonstration" given by Dr. Knudson was interrupted by the "valve stopping".

When Dr. Murray O. Hayes and I met with Dr. Knudson in your Salt Lake office and talked with him prior to the experimental test, Dr. Murray O. Hayes and I went over with Dr. Knudson accounts of several "demonstrations" that had been given before of the radiant energy device. He covered in detail the one given to Dr. Harvey Fletcher and to Dr. Carl Eyring and others. I asked Dr. Knudson if any of the explanations of the theory I had given him were other than scientifically and electrically sound and correct. Dr. Knudson said they were not; all was sound and correct; that everything I had told him was okay. I asked Dr. Knudson if I were to give him the same kind of a "demonstration" as all the letters he had read on radiant energy which described what had been done in the past, if he would also give me a written statement. I told him that was the understanding under which I would be willing to give him the opportunity of witnessing this experiment. It had been on this basis that I had talked this test over with Mr. D. V. Farnsworth and Dr. Hayes when they and you, Mr. Richards, requested I give my consent to Dr. Knudson being present at a R. E. test. This, Doctor Knudson agreed to. I told Dr. Knudson I expected to run an experimental test of the radiant energy equipment right away, and that we might as well run it in a day or so as later. I told Dr. Knudson it would only be upon this basis that I would consent to give the "demonstration" with him present. I had asked Mr. Farnsworth and the others before I met Dr. Knudson what good was to be accomplished by "demonstrating" to these doctors of science. I said they come,

Fig. 49

Memo on Dr. Vern Knudsen

they see, they believe, but dare not speak of what they see and least of all, do they dare to write the details of the tests. Dr. Murray O. Hayes said, "Dr. Knudson is different, he is afraid of no ones opinion. He will give you a letter."

I told Dr. Knudson while in your office just what I would do and what I expected of him. He said if I showed him the things recited in the other accounts which he had read, he would be satisfied and okay the device as being all I claimed for it. I showed him all of these things and more. The fact that he could not continue his "tinkering" was his fault. However, the scientific investigation had ended, and so what happened should not stop him from telling and writing about what he saw and heard. Nothing more, nothing less could be expected. He promised he would give me his opinion in writing. After the test, when I asked him if he could see any way that I could have "faked" the "demonstration", he stated most emphatically that it could not have been faked, repeating that there were no indications of faking whatsoever; that such a demonstration could not be faked. Later, after he had examined the transformer of the device, he stated that with the amount of amperage he had seen taken from the device was positive proof that the current developed by or in the machine was different than in any in use today. Because the transformer would have burned up with normal current, yet the transformer showed no signs of even ever having been warm. He was informed by Dr. Hayes that the transformer had been in use under the same type of strain for many tests in the past.

It is true, Dr. Knudson said after the valve burned out, that the tests which he wished to perform but which he could not now that the device was inoperative would not have proven anything. I said, you will agree if the machine would have performed or would not have performed this would have proven nothing. His answer was, such tests would prove nothing. It was Dr. Knudson's fault that the detector burned out because of the excessive inductive kickback he subjected the device to by throwing the main switch in and out so fast and often that this continual breaking of the circuit built up an inductive kickback, which would "blow" any electrical circuit. Dr. Knudson later said to Dr. Hayes when asked what he expected to prove by his idea of these tests, that, such tests as he had in mind would prove nothing except if Moray were telling the truth when he said that the radiant energy current would not operate a conventional motor. For arguments sake, let us grant this type of current would work a standard motor. What would that prove? Such a test would prove nothing at all which would offset the scientific proof of operation that Dr. Knudson had seen. The question now arises, was this a scientific test of a scientist or was Dr. Knudson playing detective. All my theory could be incorrect, but that would not prove or disprove the various tests as to where this type of energy was coming from. Does Dr. Knudson want to

Fig. 49

Memo on Vern Knudsen *(continued)*

acknowledge he was not capable of investigating by scientific tests or to recognize the difference between high frequency, A.C. or D.C. current? Dr. Knudson acknowledged the current to be different when it did not burn up the transformer under such a heavy current load. Dr. Knudson has more than enough scientific knowledge to handle the situation so why would he not stay with the scientific facts? No one even could hint that Dr. Knudson is not qualified, therefore, the only other answer is he was afraid to stand by what he promised and saw and wanted a way out from his promise to write a letter. Another case of history repeating itself. What I can't understand is why some men have so little confidence in themselves so they let fear seal their lips. Fear of being criticized for witnessing a new phenomena. Dr. Knudson, like all the other men of science who have seen R. E. could not point out one little thing in the whole test which was not scientifically sound and correct.

One might now go into the matter of how Dr. Carl Eyring changed in his attitude because fear of criticism made him make foolish statements rather than acknowledge facts and stand by his original promises and his statements that all Moray had told and shown him was scientifically sound and correct. So Dr. Eyring started to argue with Moray at the B.Y.U. and came up with his induction from power line theory. He argued that a 100-watt standard light globe required three amperes of current to operate. This was the same as saying that 100 divided by 115 equalled three. Dr. Eyring is a brilliant scientist as also were many of the others, but fear makes us humans say and do strange things. Just another case of history repeating itself.

Dr. Harvey Fletcher gave as his theory that the energy that was being received by the R.E. device was from some form of battery reactions in the tubes which were kept alive by the action of the universe. He stated, however, that if this were the case, the results were far more wonderful than the theory which was claimed by Moray. Dr. Fletcher said the current was high frequency. Dr. Knudson also admitted that the current received by the Moray device was high frequency.

I never consented at the meeting with Dr. Knudson in your office to permit the device to be "tinkered" with. It was Dr. Knudson's "tinkering" that threw the device out of balance by his causing the inductive kickbacks and by placing that extra coil, which he furnished, in the circuit which caused the surges and caused all the damage. This, Dr. Knudson acknowledged after he had seen the inside of the device. I tried to tell him at the first when he wanted to insert the coil in the circuit that this could happen, but he would not listen. I remember saying, "Dr. Knudson, there is no use connecting such a coil in the circuit as you will understand after you see the inside of the device. It cannot possibly prove anything and may do damage." Some days later,

Fig. 49

Memo on Vern Knudsen *(continued)*

Dr. Knudson conceived the idea and made the statement that I received the energy from an outside source, and I had a hidden wire in the 90 foot antenna pole. Please analyze this statement, Mr. Richards, This shows Dr. Knudson lost his head, and such a statement is beyond all reason. In the first place, what good would a hidden wire do? If you analyze the type of tests that were made before Dr. Knudson, what good would hidden wires do? Also, where would I get a bit 90 feet long or a device capable of operating a bit of that length to make a hole in which to hide a wire in the pole? The pole has been examined many times under the closest inspection by many qualified men and State officials. Go over these tests made with any man with a knowledge of electricity and he will have to decide such a statement could be made only because of gross ignorance or because one lost his head in a desire to come up with an excuse to avoid the keeping of promises. No one could ever accuse Dr. Knudson of ignorance. The only other answer can be fear. If you, Mr. Richards, had stayed in the laboratory instead of running all over our home upstairs hunting for the method of how it could have been faked, you would have done more to satisfy yourself than by what you did in running around the house. You ran into my wife's bedroom, the bathroom, and all over the place when she was not dressed for strangers. One thing you did find out was that she was not generating the current by running on the carpet as one scientist had suggested when he, too, lost his head. Anyway, you found we had polished, hard-wood floors, no carpets to run on to generate over four thousand watts of electrical energy by.

Naturally, I am disappointed with Dr. Knudson's attitude, but the test accomplished in part some of the things I needed to determine in my efforts to improve and stabilize parts of the device. Naturally the damage to the equipment will be expensive to repair, but that is partly offset by the pleasure I received from another recognized, noted scientist having to agree that the radiant energy discovery is all I claim for it and that he could not disprove my theory on the operation of the device or point out any faults in the presentation. This, Dr. Knudson did before another scientist who witnessed all which was said and done by Dr. Knudson and myself at this test.

Will wait with pleasure meeting you again upon your next trip to Salt Lake. With kind personal regards.

Sincerely yours,

T. Henry Moray

T. Henry Moray

Fig. 49

Memo on Vern Knudsen *(continued)*

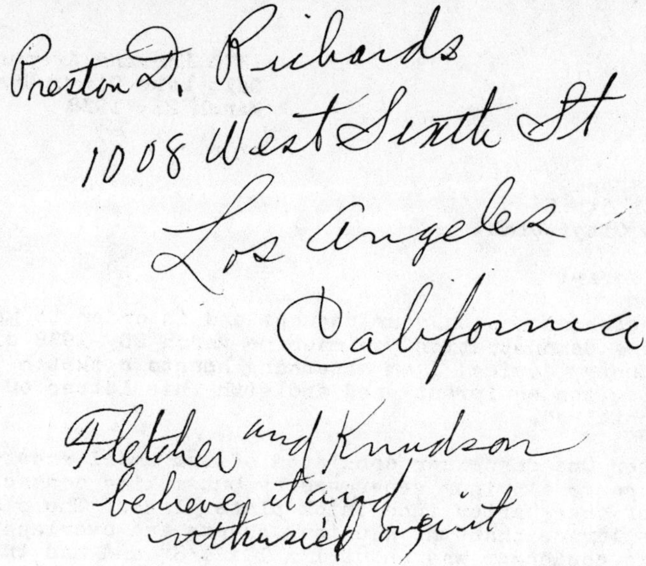
Preston D. Richards
1008 West Sixth St
Los Angeles
California
Fletcher and Knudson

Fig. 50

Memo by Moray re Fletcher and Knudsen

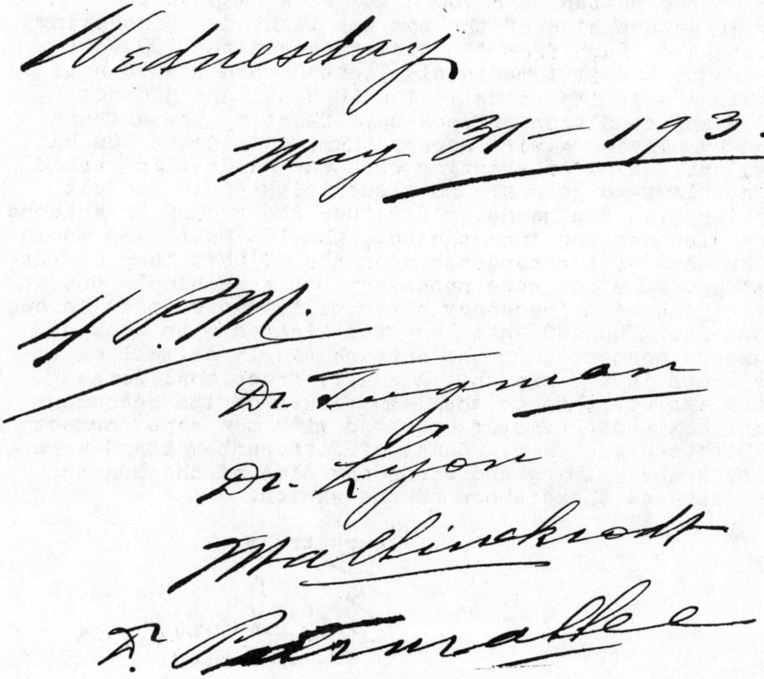
Wednesday
May 31 - 1933
4 P.M.
Dr. Tugman
Dr. Lyon
Mallinckrodt

Fig. 51

Memo on Tugman

1373 Harvard Avenue
Salt Lake City, Utah
March 21, 1938

Dr. T. H. Moray
2484 South 5th East
Salt Lake City, Utah

Dear Dr. Moray:

In accordance with your request and in order to make a record of a demonstration you made me March 20, 1938 of your Radiant Energy device, I am attaching hereto a sketch which illustrates the equipment used and with this letter outlines results obtained.

Number One condenser consisted of two small sheets of about 30 gauge aluminum separated by but making contact with a piece of one-quarter inch thick plate glass. The plate glass was larger than the aluminum sheets and overlapped them. Number Two condenser was about 1"x 1"x 1/8" and had the capacity of .025 MF. It was a commercial condenser such as is carried in stock at electrical shops and was manufactured by Igrad Condenser & Manufacturing Company.

With the 60 Watt Mazda Lamp and the two condensers attached to the antenna and the antenna and ground attached to the box containing your Radiant Energy equipment in the manner shown on the sketch, the 100 Watt Mazda Lamp on the secondary or output side of the box was lighted. Unscrewing the 60 Watt Mazda Lamp from its socket caused the 100 Watt lamp to go out, but it immediately lighted when the 60 Watt lamp was screwed in its socket. The 60 Watt lamp did not light under any condition or hook-up. Shorting the antenna and ground by placing a wire across them, caused the 100 Watt lamp to go out. Similar shorting with the hands also caused the 100 Watt lamp to go out. No electricity could be felt when shorting with the hands. If either the ground or antenna wires were disconnected from the box, the 100 Watt lamp would go out. Neither of the condensers or the 60 Watt lamp on the primary side of the box were necessary but were simply put on to show that the high frequency power will jump or pass through the condensers. The 100 Watt lamp was lighted with each of the condensers connected in the antenna singly as well as together as shown in the sketch. A small, fractional horse power motor was attached to the lamp socket on the secondary side of the box and the motor operated with the same connections used for lighting the lamp. Lamps of different Wattage were tried on both the primary and secondary side of the box and worked the same as those shown on the sketch.

Yours truly,

E. G. Jensen

E. G. Jensen

Fig. 52

Letter from E.G. Jensen

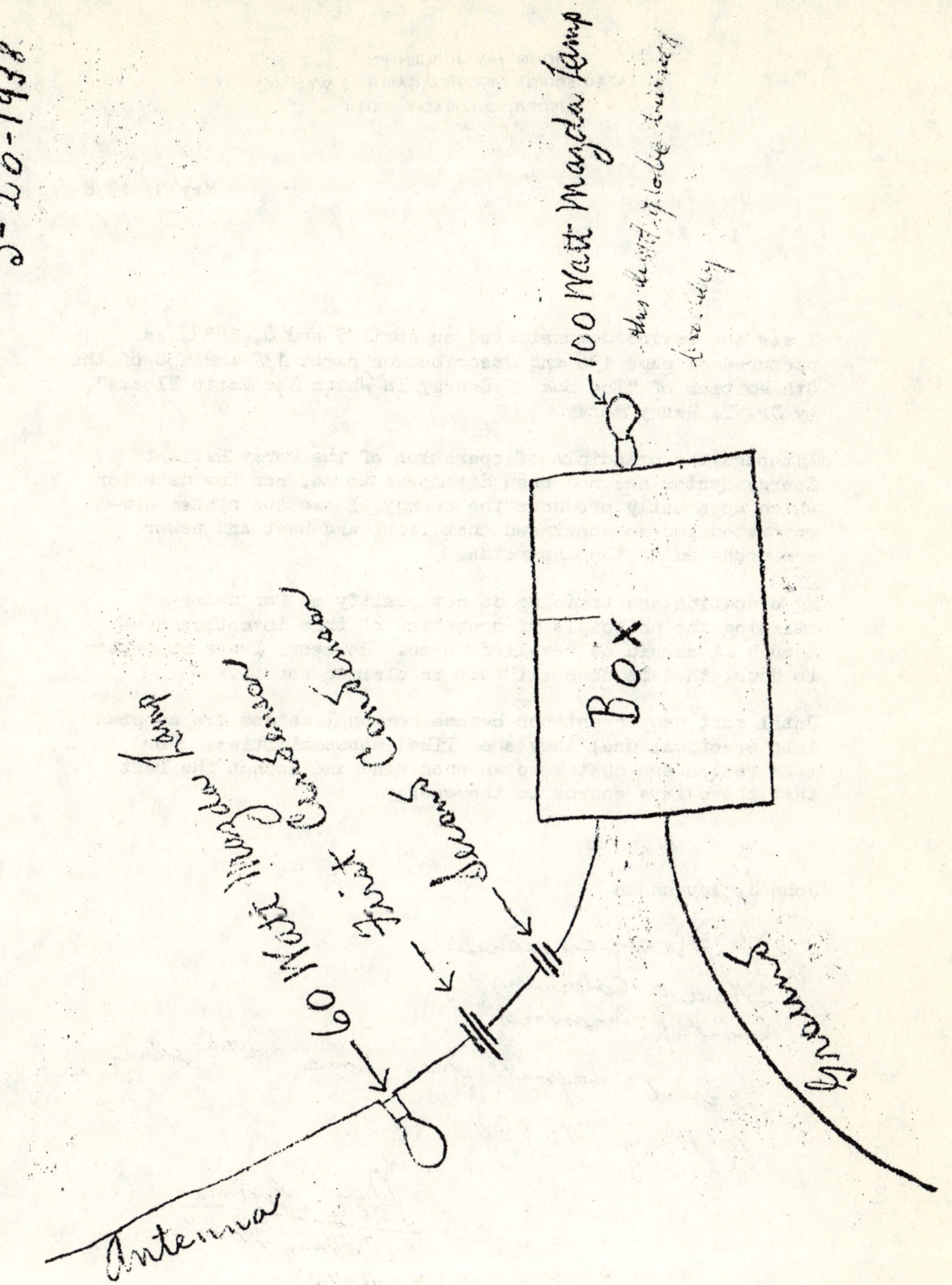

Fig. 52

Letter from E.G. Jensen *(contined)*

JOHN JAY JURGENSEN
3244 SOUTH HEATHER GARDENS WAY
AURORA, COLORADO 80014

May 3, 1978

I saw the device demonstrated on April 7 and 8, 1933, as pictured on page 136 and described on pages 137 and 138 of the 4th edition of "The Sea Of Energy In Which The Earth Floats", by Dr. T. Henry Moray.

Although the principle of operation of the Moray Radiant Energy System has not been disclosed to me, nor the detector which apparently produces the energy, I saw the system demonstrated and am convinced that light and heat and power are produced by the apparatus.

My education and training do not qualify me for understanding the principle of operation of this invention even though it should be revealed to me. However, I see no reason to doubt that it does all that is claimed for it.

Until most new inventions become commonplace and are adopted into practical use, they seem like impossibilities. For this reason one must keep an open mind and accept the fact that there is a source to the energy.

John J. Jurgensen

J J Jurgensen

State of Colorado) ss.
County of Arapahoe)

Signed before me this 8th day of May, 1978.

Patty Sebon
Notary Public

Commission Expires: 12-16-81

Fig. 53

Sworn Statement by J.J. Jurgensen

R.E. Device Used in Airplane Test

Fig. 54

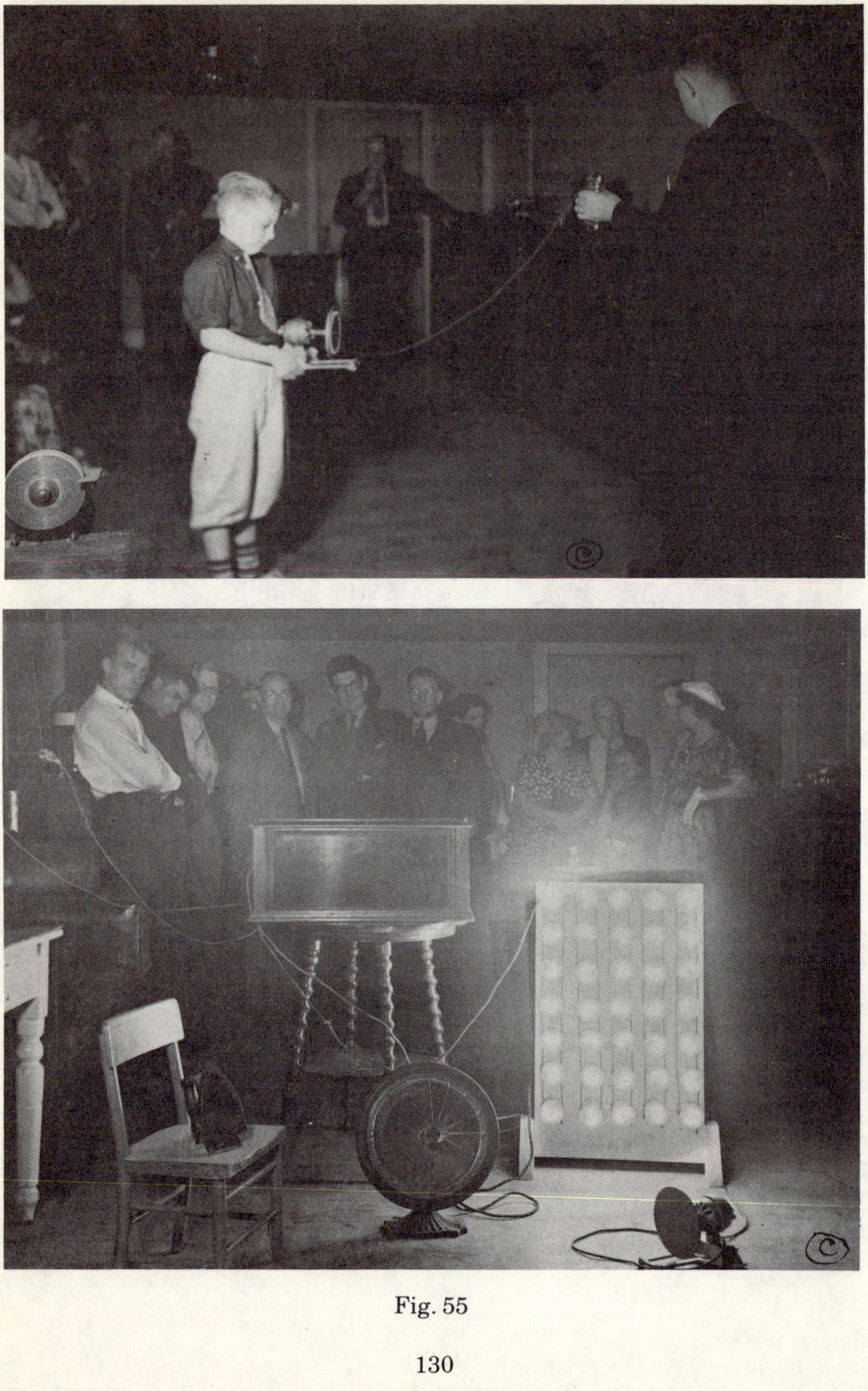

Fig. 55

Note speed of ©
Motor Going so fast
it can hardly be seen

Note dark spot around lights
Burned in film. Not actually

Fig. 55

The foregoing array of evidence should overwhelmingly document the genuineness of the Moray Radiant Energy Device. The fact that Henry Moray became somewhat of a recluse, refusing to make further disclosures to people and refusing to trust anyone in the scientific community, is more than justified when one considers the shabby treatment he consistently received. In addition, his life was often severely endangered. Further, anyone who is not naive is aware that industrial espionage is the biggest active area of espionage today. Doctors of Science have no more or no less integrity than any other group, and they are just as greatly involved in industrial espionage as are their business counterparts.

And so T. Henry Moray's Radiant Energy Device - which could have lifted every man from the dust and raised mankind to the stars - was repeatedly and methodically suppressed by greediness, suspicion, and desire for power and domination over the marketplace and over other men.

January 10, 1938
Salt Lake City, Utah

To Whom It May Concern:

I have worked in electricity all my life, was with the Utah Power and Light Company for thriteen years, and worked in all departments including sub-stations. I have been with the Kearns Corporation over seventeen years; and have charge of all the electrical work for the Tribune and Telegram Publishing Company and in Kearns and Tribune Buildings.

On December twenty-third, 1937, I witnessed a demonstration at Dr. Morays home at 2464 Fifth East of his electrical marvel box. He let me see inside this box and there was a transformer H. F. some of his cold tubes and some condensers. Dr. Moray connected this box to an arial and ground wire and lighted about thirty small 120-volt lamps. He then disconnected the ground wire from the outside and two of us held a counter poise antenna attached to glass insulators in the room and when he connected the box to this counter poise antenna I saw the same results.

During this demonstration and while the lamp and appliances were on I shorted the arial and ground wires. There was no spark, it just turned off the power from the box. I then took hold of both of these wires. There was no feeling and they were both cold.

He then took a larger box, about 18 x 30 x 15 and connected it to the outside arial and ground wire. From this box he lighted about fifty 100-watts, 120-volt lamps, ran a small motor especially wound, an ordinary electric iron, and a 500-watt glow heater.

Dr. Moray then disconnected the appliances and lamp, and attached two long wires to the box and we pulled the main line Utah Power and Light Company service switch for his home. We attached the wire from the box to the house side of the switch and he lighted his home, heated irons and heaters; in fact everything the Utah Power and Light Company service would do except run motors which I understand have to be specially wound.

Standard globes were used but they seemed to give a softer light which was more like daylight.

I am satisfied myself from my experience in electricity that there was no fake of any kind or consealed batteries. Everything was in the open so I could see every operation. I did not see the inside of the larger box.

It was a very remarkable demonstration and one which I will always remember.

Geo. R. Pyper

Geo. R. Pyper

Fig. 56

Letter from George Piper

CHAPTER 4

MORAY PRODUCTS CO.

It becomes a philosophical matter as to whether one should blame any one individual for a misunderstanding or whether one should consider human weaknesses and frailties. It is often possible to simply be in the wrong place in order to be criticized for the actions of others, even though one had no part in the actions that merit criticism.

Recently, for example, I found myself disengaging from a partnership that had become very distasteful. I am sure the other two partners believe they did what was correct. I do not doubt that justice considers well-meaning intentions and extenuating circumstances. Nevertheless, through neglect or ignorance, a person may deliberately avoid facing the implications of what he sees happening, and thus become a party to crime or serious mishap.

Our American philosophy, in a way, has become backwards. We have, in my estimation, lost the proper perspectives. What good does it do to tell a child that he has the right of way and then send him out on a busy street to walk blissfully in front of an oncoming automobile? In the Orient, a mother teaches her child that it is his responsibility not to be hit by an automobile.

I have the option in this chapter of lambasting the many promoters, lawyers and Indian Chiefs who came to Henry Moray, professing to want to help develop Radiant Energy for "mankind", proposing to try to help Henry Moray in his efforts to bring about his marvelous discovery. However, it is not my place to judge any of these men or cast aspersions on them. I am sure that in their own view they did what they thought should be done.

Recently, I had a call from Provo, Utah, from a person whom I would categorize as another crackpot. He certainly had no knowledge of my background or of the work on Radiant Energy, because his first statement was, "Can you substantiate claims that your father had an energy device?" I said, "Well, come and see," and I have heard nothing from the gentleman since. Does anyone seriously think a family spends two generations working on a scientific project just to perpetuate an illusion?

But let me develop some of the background of the business entanglements Henry became involved in.

In 1929-30, Henry Moray was offered $50,000 to sell out certain power interests, and he refused. He was approached by Yakovlev of Russia, and Yakovlev and his people insisted that Moray bring his work to Russia. Henry considered what happened to the Czar, and again he refused.

Thereafter, Henry Moray began to get threatening phone calls. He was told that there was "a contract out on him" and that he was to be killed. Is it any wonder that he did not know whom to trust?

Three men, the founders of the Moray Products Co., came to him and offered to put up a large amount of property in Mexico as a security, to sell millions of dollars worth of stock, and to help Henry Moray organize a manufacturing company, a research company, and a distribution company. Schade, Farnsworth, and Hayes offered to give Henry control of the corporation and to issue to him the largest amount of stock. Whatever he wanted them to do, they stated, they would do. Is it any wonder, then, that Henry thought that he would have control of a **bona fide** corporate operation in which no others could stop his work or threaten, as had been done previously, to withhold money as a means of forcing him to do what they wanted?

Henry withdrew from the R.L. Judd interests. Judd himself, representing Henry as his attorney, cleared the way to release Henry from his contract, although the other Judd interests took great exception to its dissolution.

Mrs. Gertrude Tracy, representing still another group, wrote Henry a letter on October 16, 1930, protesting that her people had not been given a chance to receive a fair hearing. In her letter, she referred to "your representatives" — apparently speaking of Henry's attorneys, including Judd himself. However, the letter is amusing because of the great virtue she said these people had exhibited. One is reminded of Shakespeare—"Methinks the lady doth protest too much."

"My Dear Dr. Moray:

"I can't understand your attitude in regard to our relations on the 'Radiant Energy.'

"Mr. Dustin, and his associate, came here prepared to do business in a large way. He brought to you the four requisites to that end; namely,

Unusual integrity;
Unquestioned ability of no small degree (as per references),
Experience of years in this line of organization,
Almost unlimited financial backing to carry this through to a successful conclusion.

"He was therefore entitled to a fair chance and your time while here. Instead he had nither, but a bogus contract was sprung on him and that was all he had to work with. Even then he worked out a contract (tentative) to present and that was refused reviewal by your representative.

"I don't wish you any bad luck, Dr. Moray, but I wouldn't treat an enemy as you have treated Mr. McKnight and myself and the gentlemen from California. Nor would I expect to succeed, or deserve to, after such a procedure in any undertaking.

"Mr. Todd is an ex-soldier of the World War. He has an inventive turn of mind, like yourself, and did invent something to put on tanks in airplanes (he flies) and out of which he could have made a nice piece of money, but he gave it to the Government. He came back from the War with a shattered body and has been all of these years gettingpatched up to a semblance of health. He has recently inherited several millions of dollars and was willing and eager to put some of to work for you and your inventions if you had given him and Mr. Dustin an even break.

"In my opinion you have let slip through your hands one of the most capable and strongest group of men in every way for your undertaking that you will meet up with in many a day. They measured up in every respect and would have safe-guarded your interests to the last detail. Mr. McKnight and myself brought them here and I ask for an explanation of the treatment you gave them while here.

"I gambled on you and your fairness to the last moment and lost. Mr. McKnight and I are both hurt and justly indignant at such horse play.

"You may ask how I know they are the best men for this great undertaking, a woman's intuition, coupled with Mr. Dustin's list of accomplishments and his very evident inherent integrity. Please let your wife read this and be guided by her."

On the surface, the offer Mrs. Tracy's group had given Dr. Moray was very interesting. They had an interesting idea of raising money. Payment of one million dollars was to go into the corporation. What could be wrong with that proposition? All the funds were to be withheld until three months after Henry completed the device. But if Henry had already perfected his device at that time, we ourselves would not still be trying to do research to be sure of how to standardize the "tubes." It is important for the layman to understand that although through the years various generations of the device were demonstrated, none of these devices was totally new; rather, Henry Moray cannibalized the previous version of the machine to make each new version. Also, it had come to Henry's attention that Mr. Todd was penniless, and not a millionaire as reported.

Finally, Henry's attorneys advised him against doing business with these people, since they promised nothing until three months after the unit was perfected, and then to pay only in small increments. Business cannot be done on woman's intuition or because someone is a war hero; instead, immediate funds are necessary for research, operation, capital investment and supplies.

This type of "rigged offer" seemed typical to Henry of all his negotiations through the years. Just because Henry had a "breadboard" unit that brought in energy, why would anyone knowledgeable about research and development of technical devices think that all research had been completed and that no further engineering needed to be done? Dr. Moray was no more ready to proceed to production than Bell Laboratories were when they first had a transistor device in their own laboratories, or than General Electric was the first time Tracy Hall built a press which made synthetic diamonds. It was almost fifteen years before synthetic diamonds, even of industrial grade, were finally put on the market using Hall's process.

While the era of the individual inventor is not dead, it is extremely difficult for an inventor to take his device from a laboratory conceptual model to a completely engineered, marketable, reproducible device. The complications of producing a consistent product suitable to be manufactured in great quantity have become so gigantic that no one person can possibly comprehend all the ramifications, and few have the capital outlay required for such a massive undertaking. Consequently, for anyone to demand that Henry Moray have completed and produced a device, or that Cosray Research Institute have a finished energy device ready to go on the market today — that is truly pie in the sky.

All of the offers made to Henry had limitations. However, finally three gentlemen seemed to offer him a legitimate deal. They guaranteed to raise a specified amount of money in a certain amount of time, not depending upon Henry's progress, but giving him adequate time to work on the project and ready it for marketing. This offer enticed Henry to form a company call the Moray Products Company, incorporated under the laws of the State of Nevada in the city of Reno, with the eleven original stockholders as the board of directors. A total of two million shares was authorized, of which one hundred shares were held among ten of the men and fifty thousand held by Henry Moray himself. The corporation started out vigorously and everything seemed to be going well until Henry found that a large amount of his personal stock was being sold; neither he nor the company was getting the benefit of the sales. There was some treasury stock being sold, but the millions of dollars that had been promised from sales in the United States and Mexico were not being generated. Consequently, the company was doing business hand-to-mouth.

As a result, Dr. Murray Hayes found it necessary to look for work elsewhere. He took a job with the Patent Divison of the U.S. Navy, but he agreed to continue to work on the energy patent if Henry Moray would deliver to him some of Moray's own holding in company stock. Many wild and extraneous stories began to circulate, some of which took years to unfold and come back to Henry Moray. About this time, Henry's closest friend, W.H.(Bill) Lovesy of the Utah Oil Refining Company mysteriously killed in a one-car accident. His passenger, a hitch-hiker who has never been identified, got up from the accident and walked away.

At this time Henry Moray prepared and published a booklet called **The Patent Situation,** which covers the dealings of the Moray Products Company and is included here. It also details much patent information which I will not repeat, but will refer to in the following sections.

Whenever I have asked for any direct information from the former officials of the Moray Products Company I have been given an evasive answer, in fact I was given practically no more information by them than if I had not been connected with the company.

Prior to the time that Murray O. Hayes left for Washington, D. C., I endeavored in every way to so thoroughly teach him every detail of the Radiant Energy Invention that he would know all the details, that I no longer would be the only man in a position to carry this work on. This was not done in a day, a month or a year but through a period of several years, and at no time was I able to see that Dr. Hayes was in any way betraying the confidence placed in him. He had the full detail description of the device given to him little by little, starting as far back as 1928. By October, 1929, in a letter to Mr. W. H. Lovesy, one of the leading officials from the Utah Oil Refining Co., he stated that "Moray has explained all to me without reservation and I am sure that this is a revolutionary and epoch-making invention." He even acknowledges that I had explained the detector to him and to quote from his letter in referring to the detector: "In this he has applied a fundamental principle of electric circuits which, I believe, would not be noticed unless pointed out by him. This element of his device also, as above mentioned in respect to the circuits, has numerous features which appear incidental, but are the heart of the matter and of first importance." To quote further from this letter, "I have seen the parts from which it is built. Recently Moray has shown to me the wiring diagram of the assembly and I am free to say that I can find no inconsistencies in it nor anything that does not appear to be logical and sound—and based on recognized laws of electricity when all is explained. There are many features which appear to be incidental but they are in reality of basic importance."

I want to here point out a few things. You will notice Dr. Hayes speaks of features which appear to be "incidental" but are of basic importance. He states that these things would not be noticed unless pointed out by Moray. I will refer to this statement of Dr. Hayes further on.

Dr. Hayes has told many men that he knew enough about the Moray device so that he could make one of the machines himself from the information Moray had given him. In confirmation of this statement see Mr. Lovesy's letter of June 29, 1932. Dr. Hayes also made the same statement to Mr. S. E. Bringhurst of Murray, Utah, and others.

In a letter addressed by Mr. W. H. Lovesy to Murray O. Hayes dated May 7, 1930 (and it must be remembered that Mr. Lovesy and Dr. Hayes were very, very intimate in regards to the Moray inventions) it is plainly shown that Dr. Hayes had told Mr. Lovesy that he was working on the patents of the Radiant Energy Device. To quote from this letter "I think you have memorandums prepared and possibly the greater part outlined of a patent presentation, and I think you should prepare for Henry a write-up, as near as you feel it possible, to cover the entire proposition thoroughly and completely, so that he will have it in his possession for use if desired, and it may be required at any time."

I merely quote these things to substantiate my statement that Murray Hayes was given every opportunity to know. It must be remembered that Dr. Hayes represented himself to be and his records showed him to be a patent attorney of the highest grade. In addition, Dr. Hayes was a physicist of no mean ability. He had received his A. B. degree with Major in Physics. His Masters Degree included graduate courses in physics and mathematics. He had been acting head of the Department of Physics of the Brigham Young University. He had spent five years in examining corps U. S. Patent Office. He was a man of Science, one who should know, one who does know Science. A man that when he has made the statements that his original letters in Moray's possession show that he has made relative to this Radiant Energy Machine, it leaves no loop hole for him to later say I did not know, I did not understand, I have not been shown all.

Prior to the time that application was made Murray Hayes had everything in his possession. He selected, without any advice from Moray whatsoever, the man whom he wanted to make the patent drawing. The drawings were checked, the specifications were checked, he was satisfied. Moray was not trained in patent requirements; he had only Dr. Hayes to depend on that everything was as it should be. Dr. Hayes went to Washington, not as it is generally supposed to prosecute the Moray patents, but to accept the position in the Patent Division of the U. S. Navy, because his co-partners who had agreed to put the Moray Products Co. over had failed financially to do so.

In order to prove my last statement I will again refer to letters in my possession. This I feel is necessary, not to make men's private affairs public but in order to prove that even in their sworn statements in the Second Judicial District Court, State of Nevada, in and for the County of Washoe, when they instituted restraining and enjoining proceedings against Moray under date of May 2, 1932, No. 39982 (Summons) to quote page 4, "That said Murray O. Hayes spent considerable time in the city of Washington, D. C. and is still in said city of Washington prosecuting the application for said patent," they disagree with the facts in the case. Murray Hayes took the position as stated above because he financially had to do so. And why, because those who were supposed to raise sufficient money to carry on the work of the Moray inventions and patents had failed to do so.

To quote from a letter received from Murray Hayes Jan. 15, 1932, "The cost of sending our boys to school here is terrific, except that they are in school we are no better off here than we were in Salt Lake, but they are going ahead, and that is the important thing. They could not possibly have gone had we not moved." To further show that Murray Hayes was disappointed in the results obtained by Mr. Schade and Mr. Farnsworth, let me quote from his letter of Jan. 9, 1932. "I, too, think that D. V. should make good on his Mexican proposition." (By D. V. he is referring to D. V. Farnsworth and his Mexican proposition was that Mr. Farnsworth had made so many rosy promises as to the millions of dollars that he would be able to raise in Mexico.) Now to quote further from Dr. Hayes' letter to Moray of Jan. 9, 1932: "If Knight can sell his own stock F. and S. (Farnsworth and Schade) should be able to move some if they buckled into the harness.—Why can't they do the same for Moray Products?"

Again in a letter from Hayes to Moray dated Jan. 23, 1932, "I sent D. V. some time ago, a letter, asking why he does not try Mexico, but haven't had a reply as yet. Have not had a letter from him for a long time, **because there is no progress to report I suppose.**"

I have another letter from Dr. Hayes in which he tells me I better get a job, for the Moray Products Co. would never be able to support me. I will come back to these statements later and show that it was because of these statements that I considered it necessary for me personally to endeavor to do something to raise the necessary finances to help carry on my inventions. It has been complained that I sent men to Reno to sell my personal stock in competition to the men who were trying to help me put things over, but I can prove by their own statements in my possession that they were failing, and that my only purpose was to lend a helping hand. As far back as March 16, 1931, C. Fred Schade, in writing from the Golden Hotel at Reno, Nevada, to Moray had this to say, "We expect to leave here in a few days for the coast, but will keep Murray posted (you will note they say they will keep Murray posted, that was always their attitude, they never kept me posted). Things are going slow here now, and we feel it is a waste of time to remain here any longer." Does this not plainly show that these men were through with Reno. How then if I worked a field that they were discarding was I running competition to their efforts. When they were failing to raise necessary funds, how was I hindering by trying to use my own efforts to carry on.

I have been accused of withholding progress. I have not withheld any progress, but how could progress be made without the finances necessary for such progress. They had agreed to furnish sufficient sums to secure Moray a laboratory to carry on his work. See the Assignment made by Moray to the Moray Products Co. at the time of its organization. "It is further understood that the undersigned T. Henry (T. H.) Moray shall retain personal ownership and exclusive charge and control of the laboratories and equipment used by him and his assistants for experimental and development purposes." If the men I had counted on to furnish me this laboratory could not do so, should I lay back and not try to do something myself? They acknowledged that in this they also failed as I will show by letters received. But first let me state that they had agreed to not only furnish me with a laboratory but also to pay me a salary of $1,000.00 per month as soon as the company was financially able to do so. That this $1,000.00 per month should date back from the time that the company was organized and until this $1,000.00 per month could be paid, the Moray Products Co. should pay me sufficient to support myself and family and to pay my indebtedness. Was this ever done? No. The money that Moray did receive these men sold his personal stock and after subtracting their commission of at first 33 1/3 per cent and later 20 per cent which they kept personally, Moray was given the balance. They may have changed the books and records to show otherwise, of this I do not know, but I can positively prove by written evidence in my possession that it was Moray's personal stock that was being sold.

Not being able to perform the job that they had agreed to do, they began to look for some place to unload the responsibility of their failure. In a letter written by Dr. Hayes to Moray, Nov. 13, 1931, to quote, "Last Thursday night I went to a meeting of the IRE and got talking with a man who runs a Radio and Television Laboratory. He says that with the device as far along as it is he can get it ready for market in three months if we will turn him loose on it. He will not require any pay until it is ready except some stock.

"From what I can learn of him he is thoroughly reliable."

In answer to this letter I wrote Dr. Hayes asking him why it was that his attitude had so thoroughly changed that his advice was contrary now to everything he had said before. (Dr. Hayes knew of the many chances I have had that were far superior to the offers they were now making me. That he had advised me strongly not to accept any of these offers knowing that they were contrary to the very thing that I had worked so hard to accomplish, i.e., that I should have my own laboratory in which to perfect my own inventions.) In this letter I asked him why it was that he was now willing that I should turn a man loose on it or in other words disclose all my secrets to a man that all he could say about him was that he had "got talking with," not even giving me his name; not personally knowing the man but only from what he could learn of him he was thoroughly reliable.

Moray asked Dr. Hayes to give him the man's name that some check might be made to ascertain the reliability of this chance acquaintance. No answer was ever made to these questions, but in answer to the question by Moray to Hayes as to why he had made this request to turn this man loose on it, Hayes made the reply that it was simply a case of not having the money to do it ourselves and that he thought it a good thing when they could not furnish me with a laboratory to let somebody try who did have a laboratory. I have later checked and as near as I can find out from what Mr. Farnsworth told me on his return from the East, the laboratory referred to was John Wolf's laboratory.

In talking to men who have been in Mr. Wolf's laboratory they tell me that his laboratory is no better equipped for my purposes than the one I have.

Dec. 20, 1931, D. V. Farnsworth, writing Moray from Brooklyn, N. Y., requested Moray to permit a commercial tube factory to build his tubes, stated "They make so many they do not know or care what they are for. There would be no chance of anyone getting next to what you even wanted them for." Any reasonable man can see that this was one of the best ways for Moray to betray his stockholders and a fine way for him to throw away all that he had worked so many years to accomplish. For it must be remembered that we have no patent protection until the patent is granted sufficient to warrant such action. And furthermore, these requests were contrary to the very reason why Moray had secured the services of these men, i.e., that he should have a laboratory in which he could accomplish the perfecting of his inventions.

Before leaving this subject I would like to be plainly understood that Murray Hayes knew from the beginning of my acquaintance with him years ago and I can so prove by those acquainted with my relationships with Dr. Hayes that he thoroughly understood what my hopes and ambitions were and the things for which I stood and had worked so hard to accomplish. When Mr. Farnsworth and Mr. Schade came into the picture I again made myself clear that I was not interested and would not consider any proposition that did not guarantee to me, among other things, the following:

Sufficient finances to provide myself with a laboratory and the necessary equipment wherein my inventions might be perfected under my supervision in my own laboratory.

That during the time I was engaged in giving my services for the advancement of my inventions, that those dependent upon me should be properly cared for, and that any attempt to curtail, supress or "salt down" any of my inventions when I should deem them perfected and ready for commercial use should constitute a breach of contract. Or in other words I was determined that the public should be permitted to enjoy at a reasonable cost the benefits of my inventions.

Now to quote again from the above mentioned Summons No. 39982 they further state "that the defendant herein, Thomas Henry Moray, has refused, neglected, and does now refuse and neglect to furnish plaintiff herein with models, drawings and demonstrations which the patent office requires for the perfection of the patents on the machines and processes referred to in the contract entered into by plaintiff and defendant on or about the 31st day of Jan. 1931, although this plaintiff and its attorney, Murray O. Hayes, have on numerous occasions within the last year written to said Thomas Henry Moray and demanding that he furnish the necessary information plans and demonstrations required to perfect said patents. That said defendant in answer to all these letters has refused to comply with the terms of his agreement with plaintiff herein."

In answer to the above I will state that no one can show any letters where at any time prior to the beginning of court action have I refused to render all the help possible in the prosecution of the patent application I have in my possession every question that Murray Hayes asked me and I also have in my possession copies of the letters that I sent to Murray Hayes, and in these letters I always stated that if I had not made myself clear for Hayes to tell me so and I would be willing to go over the whole matter with him again. It is true I expressed my disappointment and surprise at Mr. Hayes' seemingly ignorance of the device after my years of association with him, in which I had tried hard to fully acquaint him with all details of the Radiant Energy Device. On December 21, 1931, Mr. Hayes wrote me among other things and referring to one of my letters to him in answer to questions he had asked "The information came through all right. I believe that is enough to make full and complete disclosures, so everything will be O. K." Soon after that letter Dr. Hayes again asked me for more information. I should not say more information, for he was asking the same questions over and over again in a different way—questions he had asked in the beginning of my disclosures to him while he was still in Salt Lake City years before—questions that I had answered him in detail in which I had copies in his own handwriting of my answers to these questions.

I wrote Mr. Hayes informing him that I had given him the same information time and again, expressing my regret that I was not in Washington so that I might again go over with him every detail, telling him if necessary to pull things out of the Patent Office, delay the action until I could get sufficient money to come to Washington. In answer to my expressing a desire to come to Washington, he told me such an action on my part was not necessary, and to quote from a letter from him he states: "It would be a fruitless waste of money for you to come to Washington. Our attorneys are fully qualified to do all the argument necessary."

It is true in some of his later letters he asked me to come to Washington but that only to demonstrate the device, plainly informing me, however, not to come unless I came prepared to demonstrate. This Mr. Hayes positively knew I could not do, and I will later take up the question of demonstration.

I would like to quote from another letter written by Murray Hayes March 24, 1932. In writing to a certain gentleman who made an inquiry regarding the patent: "The patent situation on the Moray invention is very good. I had a talk with the chief of the division where the Radiant Energy application is being examined, and he was favorably inclined. I cannot personally do the patent work, being in the government service, but we have some of the best men in Washington handling it. Do not tell Moray what I have told you, as we are trying to get him to do some things that will mean rapid progress for the company, but he probably will not if he learns that things are going well here. If we can get him back here to demonstrate it will mean all the money we need, but he will not unless he believes it necessary to save things. The sooner we can get him back here to demonstrate, the sooner there will be returns from the company."

I would like to analyse some of the statements in the above quotation. If I had been refusing any of the requests of the patent office, how could the patent situation be "very good?" Why should there be that "don't tell Moray?" Does not that show that they had been trying to deceive me? What were some of these things they were trying to get Moray to do?

They were, as shown by former quotations, to give my secrets to strangers, not to the Patent Office. They already had everything. Murray Hayes had not been asking for this additonal information for the Patent Office. He had been asking it that he might turn this information over to those that they later acknowledged had tried to make my machine (and that contrary to all agreements. See assignment Moray Products Company) and could not make it work.

Why could they not make it work? I will give my version of that later on.

Why did they wish these demonstrations? Because the Patent Office had asked for them? No. There were other parties in the east who were saying that if I would come forward and make such a demonstration they would buy this sales and manufacturing contract that Farnsworth, Schade and Hayes claimed to have, and which was obtained by misrepresentation, deceit and the attempt to manipulate so as to get something for nothing.

Was it their aim and purpose that the Moray Products Company would receive the full benefit from this contract? No. The sales and manufacturing contract is a proposition in and of itself—a personal affair between myself and Farnsworth, Schade and Hayes. It was not their idea that the stockholders, the investors who had put their money into this thing should receive the benefit, but it was an idea of theirs that through the manipulation of words and misrepresentation that they should claim they had paid me $15,000.00 for this contract, and thereby be able to satisfy their own personal greed, and their desire to profit by something they had never earned. Let me go on record that while it is not my purpose in this account to even consider the sales and manufacturing contract, for your information I will say that the parties above referred to have not paid $15,000.00 for any contracts—they have not paid fifteen cents nor one cent for this contract nor any other contract, and I have written evidence to prove this statement. The sales and manufacturing contract has not, is not and never shall be recognized by me. If and when any profits are made from any source whatsoever on my inventions, the investors, not the manipulators, shall realize their just and right pro rata share of these profits.

I would like to call attention again to that letter quoted above written by Mr. Hayes on March 24, 1932, in which he states that he could not personally do the patent work, being in the government service. Does this statement agree with their sworn statement presented to the Court in the State of Nevada that said Murray O. Hayes spent considerable time in the city of Washington, D. C. and is still in said city of Washington prosecuting the application of said patent?

Murray Hayes took 50,000 shares of my personal stock which was to pay him for doing the patent work. He takes a job with the U. S. Government. He plainly stated in his letter of March 24, 1932, that because of his position with the government he could not do this work, but that they had patent attorneys, some of the best in Washington, handling it. Did Dr. Hayes pay these attorneys? He had received his pay from me to do this work, he is unable to do so, so they hire Cameron, Kerkam and Sutton and pay these gentlemen large sums of money out of the Moray Products treasury to do something that Mr. Hayes had already been paid to do.

How are we going to answer these questions? One time he says he is doing it, one time he says he is not and cannot If he cannot and they have to pay some one else for doing it, why should he be paid? If he can do the patent work, why should he accept payment for his services which he is supposed to render his government and which he evidently cannot be doing if he is doing as they reported to the Court?

Under date of July 15, 1931, Murray Hayes wrote me from Washington, D. C. a letter which gave me the first idea that he would be unable to openly prosecute the patent application of my invention. I will later quote from this letter. Let me state here that at the time Dr. Hayes left Salt Lake City, he did not inform me until two days before he left that he was going. When I expressed my anxiety as to my patents and who should handle them when he had accepted a position with the government, he assured me that that had been taken care of and that the U. S. Navy was so anxious to get his services and had been trying to do so for a whole year, that they were willing to permit him to continue on with my applications because he had explained to them his situation.

In the above referred to letter of July 15, 1931, in referring to the fact that he could not do my patent work I will quote: "There is a man working where I do who says he makes more from his private patent practice than his salary. He does it through an associate, but we intend to have associates anyhow, so that is O. K. by us. This is strictly confidential and especially should be kept from anyone connected with Smoot's office. I will be right on the job looking after our stuff."

Murray Hayes has from the beginning been fully aware of the danger that existed with anything as big as this Radiant Energy Device. In a letter received by Dr. Hayes April 17, 1931, it proves that he was aware of my fears of this thing being stolen from me if I disclosed too much to strangers until I had full and complete patent protection and financial backing, as many a good thing had been lost for the want of sufficient caution in handling it. This quotation that I am about to make from the above mentioned letter expresses the identical idea that is expressed in a letter written by C. Fred Schade to Murray Hayes at a later date relative to his (Schade's) conversation with a certain influential man in California.

Now to quote from the above mentioned letter of April 17, 1931: "One of the parties interested formerly held a confidential position with the General Electric and later with the Westinghouse. He nearly took my breath when I told him regarding Moray's fears that the proposition might be stolen from the Patent Office. He said, 'It will, just as sure as you send it there, that the U. S. Patent Office is honey-combed with employees of the General Motors, General Electric and other large companies. That he had helped steal valuable data from the Patent Office at the request of the above companies. He said you were crazy if you sent a description of this device to Washington before you had plenty of money to follow it through and influence enough to prevent a theft.' So I confess after leaving him that I was in error scoffing at the fears of Moray."

It was because of such statements and the infringement cases I had known of where inventors had lost all or had to follow through long years of court action in order to get what was justly theirs, that has made me so cautious, and the reason why I was so insistent that my patents should first fall directly into the hands of the Commissioner of Patents, knowing that if such an official should be the first one to receive them, then there would be no later danger of someone getting possession of my secrets who had no right to do so.

I have been very cautious, have tried hard to make no mistakes and I am fully convinced that my disclosures have been fully protected and will be protected from such things as referred to above by the Commissioner of Patents. I am a great believer in keeping the barn door closed before the horse is gone, rather than after.

Reference was made above to D. V. Farnsworth's Mexican proposition. In leaving that subject I would like to quote from a letter dated December 21, 1931, received by me from Carlos M. Gaxiola, the Mexican scientist who so thoroughly investigated my device. "Dear Sir: I have received many days ago your pamphlets related to your wonderful invention which I personally inspected in your home last year in the company of Hon. Milton H. Welling, Secretary of State there.

"I am still interested in the future of this invention and its industrialization, and would like to know if you will, as you and Mr. Farnsworth offered me then to keep me in mind for the propaganda and territory in Mexico."

Let me state here I did my uttermost to get Mr. Farnsworth to make good on his Mexican proposition but which he has entirely failed to do.

Let me now briefly cover the patent situation.

As has been stated above I sincerely tried to fully advise Murray Hayes on all details of the Radiant Energy Device, and he in turn informed me and others that he understood all and that I had shown him all. I have shown Dr. Hayes' qualificatons which prove him to be a competent scientist and one in a position to know what he was talking about.

After Dr. Hayes left here to go to Washington, the idea came to me that it would be well to seggregate or separate certain parts of the patent. Accordingly, I wrote him to this effect. In his answer to me dated July 15, 1931, is acknowledged the above statement. "We shall put in by amendment those things you mention that are not claimed. I wanted to cover everything broadly, but did not want to run over twenty claims as it costs one dollar per claim for all over twenty when filing."

I naturally took it for granted that the matter had been taken care of until a paper copy of the Patent Office report under date of February 16, 1932, asked for the division of my patent into three divisions. In other words, in July, 1931, they promised to do this thing, by February they had not done so, and as late as July 20, 1932, when I received a communication direct from the Patent Office, wherein they again ask for this same division into three classes, stating that the requirement is repeated and made final. It seems that some sort of an answer to the request of the Patent Office was made by them on March 4, 1932, but that the request of the Patent Office had not been complied with or why should the Patent Office have to repeat its request July 20, and state "The requirement is repeated and made final."

In answer to this communication I wrote the Patent Office asking them just exactly what I should do and stating to them that I was most anxious to have the papers exactly as per requirements and also asking the Patent Office if they would kindly give me information and advice so that I might place the papers in shape so as to meet the requirements of the patent office. I sent this letter air mail and attached an envelope with air mail stamps so that I might receive a quick answer to my inquiry.

The above is all the information I have as to the patent itself. Nothing is asked for in the shape of additional information, demonstration or anything but the division of the patent as stated. I have had other correspondence with the Patent Office during thè months of June, July and August in which it was arranged that all future communications relative to the patents would be sent direct to Moray.

It has been necessary for me to file with the Patent Office certified copies of the Court Decrees in order to establish my right to have this information because of the action of D. V. Farnsworth and the others. On August 10th, 1932, I was informed that under date of July 23, the Patent Office received a paper from D. V. Farnsworth, as President of the Moray Products Company, requesting that Moray be denied access to his applications. This paper was supported by an affidavit signed by D. V. Farnsworth that he was President of the Moray Products Company. It was also signed by Murray O. Hayes as Secretary and Treasurer.

There can be only one conclusion drawn from the action of these men. They were trying to get possesion of certain papers and information by the same methods that they have used in all their transactions. It must be remembered that District Judge Ben F. Curler under date of June 11, 1932, removed Farnsworth, Schade and Hayes as directors of the Moray Products Company. Then it was found that the Court had been deceived and that Mr. Farnsworth had resigned prior to the time that judgment was rendered and because of this deception by these men to the Court, it was necessary that Judge Curler, under date of July 22, 1932, issue an amended judgment and decree in order to cover this deception.

These men could not possibly have supposed that they would remain officers of the Moray Products Company in spite of the action of the Nevada District Court. Their only hope could be to cause more temporary delay and to carry on their plan of deception. Needless to say, proper steps have been taken to overcome this deception as in all other cases.

I did not want trouble. It will be noted that during my visit to Reno in March, 1932, I made no reference to any of these facts above referred to, and everything that happened prior to my visit in Reno I was fully aware of, as much so then as I am now, excepting that I was assured at all times that these men whom I had entrusted as officers of the Moray Products Company would, as soon as possible, issue certificates of stock to cover the sales that I had made in Nevada. In fact less than an hour before I left Salt Lake City, D. V. Farnsworth called me up on the telephone and assured me that the certificates would be issued without further delay. It is a matter of record that I sent to these officials a stock certificate No. 285 for 16,667 shares, and a list of the investors to whom I wished certificates issued. This certificate was the smallest one I had, hence the necessity of turning so large a certificate over to them, it covering about sixteen times more stock than that which was to be transferred.

They kept this certificate and yet did not issue the stock according to my order. They even went so far as to try later to deny having received this certificate, but I had registered it to them through my attorney, Mr. Musser, at Salt Lake and we had positive proof that the certificate was in their possession.

It is also a known fact that when Mr. Schade was asked why he had not issued the certificates to the people in Nevada from Moray's personal stock his answer was that attorney R. L. Judd had put in a restraining order against his issuing or transferring any stock owned by Moray. When Moray heard this he knew that it was just another of Mr. Schade's misrepresentations, because he knew that attorney Judd was a friend of the Moray inventions, would not if he could and could not if he would have issued such an order; that Mr. Judd's only attitude to the Moray Inventions was one of helpfulness and nothing else.

However, so that Moray could say that he had personally interviewed attorney Judd, Moray called on him and found things as he expected and that Mr. Judd knew nothing of such a claim and was surprised at the statement. Later, on about the 10th of April, 1932, Messrs. Penny and Lilinquist, unbeknown to Moray, called on Mr. Judd and Mr. Judd told them most emphatically that he had not done this thing and that he would have absolutely no reason for doing such a thing.

At last when the officers of the Moray Products Company found they could no longer beat about the bush, they instructed the Resident Agent of the Moray Products Company at Reno, Nevada, to inform me that among other reasons why they would not issue stock to my order was that the stock issued to Moray was invalid, or, in other words, that I no longer had any stock in the Moray Products Company. They were putting me out of my own company, or in other words, it was going to be a case that the creator of these inventions, should have to step down and out after having spent his whole life on this work in favor of some promoters who have never invested one cent of their own money in this thing, but on the other hand had taken out large commissions, issued themselves, because Moray trusted them, 150,000 shares of stock, from Moray's own personal block, and then made the claim that **they would either run and control the whole thing or ruin it all.**

Moray was up against the wall, he must go to court. He had tried to bear everything he could in order to protect the Moray Products Company and its investors from the results of law suits, but he had no choice, they had forced him to take an issue.

You that were at the trial at Reno know how much mud throwing was done by the other side. I was financially unable to be present at the trial because it took all my money to furnish myself with attorneys. I would like to refer you to a statement made by Moray at the directors meeting held in Salt Lake City, Friday, July 29, 1932.

I can see that it will be impossible for me to cover even the ground I had expected to do. I have been continually putting letters aside that I had intended to quote from, but this is already far longer than I intended to make it. I could show you Mr. Farnsworth's records as an oil stock promoter. I could show you what those who are supposed to know Mr. Schade in Ogden, Utah, and other places have to say about him. But if you care to investigate these things yourself, I do not think you will have any trouble in finding out what you may care to discover.

I have before me written proof that the fifty thousand shares which C. F. Schade claims to own has never been paid for. Mr. Schade under oath in the courts of

Salt Lake City, in his wife's divorce case, testified that he could not make a division of this Moray Products Company stock that he was reported as owning, because it was not his, he had not earned it. I have before me a memorandum in Mr. Schade's own hand writing acknowledging seven certificates, giving me the number of these certificates and the number of shares totaling 2500 shares, which he states "given Schade to pay some obligations with." Here Schade acknowledges that he paid obligations with stock that he wished to "term" that Moray had given him, and he states they were used to pay some obligations, yet when Moray pays obligations with Moray's personal stock, obligations incurred in the forwarding of his inventions, some of you people cry "crook." Moray paying debts incurred that the inventions might go on, out of his own personal property, and instead of this being appreciated he has been censured for it. More may be said about Moray's "crookedness" later on.

I have shown above that Dr. Hayes' claim on personal stock is based on the promise that he would do the patent work. He himself acknowledges he cannot do it, therefore he is trying to claim property without paying for the same. Mr. Schade has acknowledged he has not paid for his and Mr. Farnsworth's claim on stock is exactly in the same status as Mr. Schade's. Both were working under the same agreements, the same promises and identical conditions.

Mr. Schade, under date of October 5, 1931, went to an attorney at Salt Lake City, named E. D. Hatch, and there obtained 29 receipts of personal transactions of Moray's and some of his relations, under the false pretense to this attorney that they were for the use and benefit of T. H. Moray. This was for the sole purpose of again fraudulently deceiving the public, and he fraudulently deceived attorney Hatch when he signed the receipt that I have in my possession before me in which be claimed he was getting these receipts because I had sent him to get them for me for my use and benefit, and it was not for six or seven months later that I knew that Mr. Schade had obtained these papers and signed a receipt for them and at the time I found out this had been done I wondered what use he would try and put these receipts to. It has later become very evident that he is using them in his same old game of deception.

There has been another man by the name of H. C. Carlisle who has also made gross statements of misrepresentations. I have before me a letter dated May 22, 1932, from Greeley, Colorado, which positively proves that Mr. Carlisle on May 20, 1932, was in Greeley and there falsified and misrepresented facts so grossly that they wrote me from Greeley stating that they positively knew from their own experience that this Mr. Carlisle was not stating the facts. I will quote in part from this letter: "Mr. Carlisle told me of several instances that I knew were not so, having been in Salt Lake and knowning the facts. One especially that he knew Mr. Noffsinger was there—that they saw me while there. If so he did not recognize me while in Mr. Varvels office. He also said that Mr. T. H. Moray refused to accept papers served on him. I asked him what time he accepted the papers and he said that they tried all day to serve them and along in the evening late, Wednesday, May 11, T. H. Moray accepted them. This I know is not true because I, myself, in person, was sitting in T. H. Moray's home at 2484 South 5th East, when they were served and T. H. Moray did not refuse the papers in the least."

I could go on and show where Mr. Carlisle has come into my home and discussed matters before witnesses and gone away and misquoted and misrepresented what took place and I can show other misrepresentations made by this man.

Under date of June 13, 1932, Murray O. Hayes wrote me among other things that a certain party had come to his office to see him and that this party had told Hayes that his brother and father had been to Salt Lake City in either December, 1931, or January, 1932, and had there witnessed a demonstration of the Moray Radiant Energy Device. After some investigation I found out that Mr. Gervis Reed whom Mr. Hayes reported had made the above statement had a father by the name of Ira E. Reed and a twin brother T. V. Reed. To quote from their letter to me of June 24, 1932: "This is to clear the mistake rumored that a Mr. Reed from Greeley, Colorado has seen a demonstration in Salt Lake City or any other place." Mr. Ira E. Reed, the father of Mr. Gervis Reed, of Camden, N. J. and T. V. Reed, a twin brother, states these facts and acknowledges these statements by signing a letter of testimony.

Mr. Reed also states the fact that he nor his son have not been in Salt Lake City in the last ten years.

There has been some attempt by the former officers of the Moray Products Company to try and say that former associates with Moray have found him faulty in judgment and desirous of cheating people out of their just rights. I will quote from a letter of June 14, 1932, written by a Mr. E. G. Jensen of 903 Continental Bank Building, Salt Lake City, Utah, one of those who have assisted Moray financially in times past and who, while not having stock in the Moray Products Company is fully aware of all that has taken place; has followed Moray in his work for the past six or seven years and positively knows that when Moray realizes anything on his inventions commercially that each and every investor be they stockholders or otherwise, shall receive their just share and equity for any favor or assistance rendered Moray in his work to put over his inventions. To quote from the above mentioned letter signed by Mr. Jensen to a friend in the east:

"The more I follow this thing and know about it, the more convinced I am of the soundness of Moray's plans and judgment."

I would now like to quote from a copy before me of a release of agreement signed by certain parties to a contract of April 15, 1926. In this agreement it states that differences have arisen between the parties as to the operations of the terms of the said agreement **"for which neither party blames the other."** These people know that they shall be justly dealt with and therefore were willing to sign this release and agreement.

It has been claimed that certain parties invested $29,000.00, turned this money over to me, that I took the same, banked it in my wife's name and did not use it for the purpose for which it was given me. In the first place, they are wrong as to the amount, exaggerating as usual by nearly 100 per cent, and they are equally wrong in the claim that the money was given to me and that I banked it in my wife's name, for the very reason that the money received by these parties was at our mutual request placed in the bank under the name and check of one of their own parties. He checked through full report and invoices where every cent was to go before the checks were issued. Mr. Jensen was one of these parties and I have referred to him above. The people who put up this money are not kicking, why should untruths be hatched up by those who are in no way concerned.

The question may be asked, why did I trust Farnsworth, Schade and Hayes as I did? My criticism of most inventors has been that they have been too distrustful of mankind and they have not only wanted to be the inventor but the office boy, general manager and everything else. So I went the other extreme and gave too much power to men, so much so that they were like the camel and the Arab. Murray Hayes was the man that I counted on, the man I thought I knew after years of trying, the man that promised to stand by me if I would only give them unlimited power. I followed his advice and it was nearly my undoing. I do not intend to burden this statement with replies to their accusations of how I have misused money furnished me by the Moray Products Company for the very simple reason that all the money that I have received from Schade, Farnsworth and Hayes, was from the sale of my personal stock after they had deducted their commission. They have not lived up to their agreement that the Moray Products Company should furnish certain sums, but have even taken commissions from deals that I personally put over, in which they had no part in the transaction except to collect their commission. This was the case in the S. E. Bringhurst deal in which Mr. Bringhurst will tell you they had no part, but they did not fail to deduct their commission to cover this transaction from other transactions of mine they had made.

It stands to reason that I should put forth every possible effort to forward my inventions because I am the one that will benefit the most by the success and marketing of the Moray inventions. If it is money Moray wanted, he could have gotten that long ago. If it is the chance to put over his inventions in his own laboratory, that takes finances, and if Moray squanders the finances how can he ever hope for a successful conclusion and realization of the hopes of his life? While it is true that I have a large automobile, I had a large automobile many, many years ago. I turned this car in on this new car and have been paying for it in small monthly payments out of the small balance I have allowed myself as a means of support. No reasonable man can deny me the right to exist and support my family. No large sums have been paid out in extravagant living. I did not organize the Moray Products Company so that I could earn a living, and I have been willing to sacrifice and have done so, spending large sums of my own money and that of relations in order to carry on the advancement of my inventions, but I do not propose that my family should go in want just to satisfy the criticism of certain unreasonable, unscrupulous promotors.

It must be remembered that I have an income, and while it is small it is entirely independent in every shape and form from the Moray Products Company.

It is true that I ride horses for exercise (just as other men play golf). Sorry to say I very seldom have the chance to ride as I am kept too busy trying to make a success of my inventions. A man must keep himself physically fit in order to stand the strain I have been under, and those who object to my taking a little recreation and exercise are very unreasonable. However, I want to make the positive and emphatic statement that not one cent of money that has ever come out of Nevada has been invested in any horse of mine.

I am not going into detail on this horse question, but will say that like all the rest of the stories, it has been falsified, exaggerated and misrepresented.

I have shown above that Farnsworth, Schade and Hayes have, by their own statements, proven that they have done nothing in payment for the stock they claimed, and for the information of all concerned let me most emphatically and definitely declare that their stock manipulations in the giving away of stock and their other stock transfers is and shall not be recognized by me nor by the Moray Products Company.

There may be some anxiety felt as to just how much harm Dr. Hayes and his associates may be able to do because of the information in all its detail that Dr. Hayes acknowledges to have been given by Moray. It was brought out in the trial at Reno that Dr. Hayes and others had tried to make this machine and it did not work. That has been partially covered above. If Dr. Hayes could have done any harm, he would have done so long before now, for I have reason to believe that he has for the past eight or nine months been endeavoring to the best of his ability to betray me and my secrets, secrets that were entrusted to him in a professional way and therefore they became professional secrets. You will notice that he mentions that there are many "incidental" features or rather features that appear to be incidental and would not be noticed by anyone unless they were pointed out by Moray. These apparent incidental features are in reality of basic importance and to quote from his letter "has numerous features which appear to be incidental but are the heart of the matter and of first importance."

It is very evident that Dr. Hayes has forgotten the telephone number of these numerous incidental features that are the heart of the device, and having forgotten is unable to recall them and therefore he can put no light in the machine without the heart. Because of Dr. Hayes' faulty memory it apparently seems that I still am the only man of whom the public has any knowledge that can duplicate this Radiant Energy Device.

As to demonstrations: So many stories have been going around relative to my giving demonstrations as mentioned in a letter received from Dr. Hayes August 27, 1932: "I have first hand, written, unimpeachable evidence that you have recently been giving some of the most successful demonstrations of your career."

If it were not for the fact that I also want to repudiate their bald faced lies in this respect also, I do not deem the source of their "unimpeachable" evidence worth considering. The only direct and concrete evidence that they have been able to show is that there is one party by the name of John Burt who even told me to my face that last October, I gave him a demonstration. Mind you a private, personal demonstration, to a man who has neither the financial backing, the community prestige nor the scientific training that a demonstration would amount to anything one way or another, yet I go to the danger, the expense, the trouble and the time to give this man an absolutely useless, uncalled for and fruitless demonstration. I am not going to argue the man's sincerity, verasity nor anything else, but can say that if it were necessary I can prove by at least twelve people that this thing is not so. If the man is subject to hallucinations and great imagination I cannot say, but I would have no trouble in establishing which one of us is off.

It has been claimed that I gave a demonstration in the Y. M. M. I. A. "M" Men's Class of the Forest Dale Ward in Salt Lake City. Again to answer this foolish statement, what would I gain by such a useless demonstration? Am I still out trying to prove this thing? The Moray Radiant Energy Machine is no longer on the defensive, it is a settled, proven fact. But to answer that as I have all their other lies, I will state that I have not been inside the Forest Dale Ward buildings during the past eight years, and never at any time during my lifetime have I attended any Y. M. M. I. A. meetings in the Forest Dale Ward. The supervisor of the class, when questioned by me, stated that Mr. Carlisle had asked him concerning this purported demonstration, and he made the statement to Mr. Carlisle most emphatically, that such a demonstration had not been given in his class. The bishop of the Ward, the president of the Mutual nor any officer of the Ward knows anything of such a demonstration.

Now why should I go ahead and give a lot of these useless demonstrations, and what would be gained by so doing, and what would be my object in lying about it? If I could give demonstrations and should feel like giving demonstrations there is no need of my hiding aroung the bush and lying about it. No one has any right to say that I shall or shall not give demonstrations. I have full charge, personal ownership and control of the laboratories and equipment used by me and my assistants for experimental and development purposes, and this device most certainly is still in its experimental and development stage. Having such unlimited power why in the name of reason should I lie about it?

The device was burned up after a noted

scientist had completed his scientific investigation and could find no fault with the demonstration so began his tinkering and tried to turn police detective instead of remaining in his realm of science and he did what can be done to any electrical machine—he burned it up. A very interesting account of this demonstration may be found in two letters written by Dr. Hayes, one on May 25, 1931, and the other May 27, 1931. Since that time I have been left without sufficient substance of one particular kind to carry on further demonstrations, and all these reported demonstrations have been hatched up and are the product of a feverish mind. Their unimpeachable evidence, if hunted down, would be found to be a will-of-the-wisp.

Heresay reports have come to me that the Moray tubes are lost, that through my failure to perform certain things (and these things have never been defined to me that I have failed to do) but the accusation is that something, a missing link, some indefinite, unexplained something is missing and they seem to be unable to fully inform themselves just what it is they are talking about. I think I have proven and shown above that Murray Hayes acknowledged to have at one time known all. Then if we turn to his letter of August 7, 1931, written at Washington, D. C.: "I spent the whole day in the search room of the Patent Office and examined every patent issued by the U. S. on devices for the reception of Radiant Energy. **Most of them were so obviously dissimilar to your invention that they did not consume much time, and not one is there which shows on reading, even the remotest resemblance of your Radiant Energy Machine. Not one mentions such an idea.**

"Neither is there any patent which suggests anything like your oscillator tubes."

If at the time the patent was applied for in July, 1931, and then in August Dr. Hayes makes the above statement, how in the name of reason can the tubes be lost, if Dr. Hayes and his associates have done the right thing when we were in the patent office prior to the time of any other patent which suggests anything like Morays oscillator tubes.

If mistakes have been made, they are Dr. Hayes' mistakes. He was the patent man and should have known, and I am convinced did know, and that the patent situation is exactly as we quoted from his letter of March 24, 1932, in which he stated "the patent situation on the Moray invention is very good."

They have also made the statement that the lead and rubber discoveries are also lost. These statements, like all their others, I am fully convinced are exaggerated, misleading and falsified. How can they say that these two discoveries are lost; can they name the parties who are able to do the same thing; can they give the patent numbers of patents where the same thing is accomplished? And after they have done all of this (which I am positive they cannot do), that would still not prove their statement. Even if someone else could accomplish the identical thing that is accomplished in this lead and rubber discovery, that would not prove that they had discovered the method used by us. There would be room for two or three different methods to accomplish this same thing without greatly affecting the value of one method. How can they say that our method has been lost when they have not the slightest idea how we accomplish either the lead or rubber discovery. These two discoveries have never been disclosed to any man, and at one time when Murray Hayes asked me last December to apply for the patent on the lead and the rubber, in my letter to him of Dec. 7, 1931, I stated, "The lead and the rubber will never be patented until they are sold, and then those who buy them can do as they want and the seveneteen years will not have started. I can send samples of either any place and have no fear of anyone making them."

If I have never disclosed these two discoveries to any one, how can they then say that someone else has found the same thing?

In closing let me say that the court proceedings at Reno during the month of May, 1932, proved beyond the shadow of a doubt the genuineness of the Moray Radiant Energy Device. The hundreds of demonstrations given to some of the World's greatest scientists, both foreign and American, and the hundreds of intelligent people who have seen demonstrations as have been recorded in the accounts of these demonstrations, prove beyond a shadow of a doubt the genuineness of what is claimed.

And last, my enemies, the former officers of the Moray Products Co., have done everything, have hunted everywhere, have stooped to the lowest, most contemptible falsehoods, but through it all the Moray Radiant Energy invention has come unscratched, its genuineness proven and unimpeachable.

I appeal for whole-hearted cooperation of all the investors to the end that we may, with the least possible delay, advance to the success which will mean a realization beyond your fondest dreams and hopes of the success of this enterprise.

T. H. MORAY.

Directors meeting of the Moray Products Company, held according to notices mailed at 2484 South 5th East, Salt Lake City, Utah, Friday, July 29, 1932, at 2 P. M. Present: T. H. Moray and S. E. Bringhurst, a majority of the directors appointed by Court action of July 22, 1932, Reno, Nevada. Proof of notice to Walter Rowson, the third director appointed by the court, is contained in a telegram of July 28, 1932, from Reno, Nevada, and a copy of a letter of July 8th, and telegram of July 26, 1932, to Mr. Rowson and which will be found in the files of this office.

Meeting was called to order by T. Henry Moray. T. Henry Moray was voted upon as President of the Moray Products Company. S. E. Bringhurst was voted upon as Secretary-Treasurer and Walter Rowson was voted upon as Vice-President of the Moray Products Company.

Mr. Moray and Mr. Bringhurst executed oath of office as directors and as President and Secretary-Treasurer, respectively. Blanks were sent to Mr. Rowson at his Reno address requesting that he execute the regulation oath of office. T. Henry Moray desired and did make general statement for the benefit of stockholders and officers of the company, and it is hereby made a part of the record.

"I would like to have spread upon the minutes of this meeting some few facts that are very vital and some that are more a matter of comment.

It is a matter of common knowledge that with a few other stockholders, I petitioned the Court at Reno, Nevada, that new directors of the Moray Products Company be appointed because of the failure of the old directors and officers to comply with the laws of Nevada and the By-laws of the Moray Products Company. There were also other reasons why it was necessary that a reorganization be affected which might be summed down to the fact that no progress was being made and that time was being wasted.

No matter what they said at or about the trial at Reno,—We (those who petitioned the Court) won the decision of the Court, notwithstanding the many, many witnesses and their statements. No attempt was made by my attorneys to place any counter testimony as to my character for the simple reason that my character or reputation was not on trial. To any of those who wish to know about me personally I will gladly furnish references of reliable citizens of this State and elsewhere who have known me for more than thirty-five years and my record is such that I am willing that it should be gone into in as much detail as anyone may desire. Those who testified as to my character at Reno, strange to say, do not know me more than in a casual way and I have letters proving their testimony was based on their own purposes and not on fact and truth. I will dismiss this with the statement that I enjoy a reputation that cannot be impeached and while I feel sorry for those who for ultra motives have tried every kind of method to gain their point I feel great satisfaction in the fact that they have failed in their efforts to ruin the value of the inventions I am trying to put on the market to the benefit of mankind.

I wish to go on record as saying that nothing shall go undone by myself or those connected with me over which I have any control that will tend to promote the interests of the investors in these inventions. Every investor, large or small, shall be protected to the best of my ability. Much that I have said has been twisted and misquoted and the spirit of my attitude changed by my enemies, but investigation of both sides will show where the fault lies.

As to the patent. Much has been said by those just removed by the Court action. Personally, I know very little as to the patent situation, but by recent letters in my possession I do know that the only complaint, if it might be so called, has come through the failure of those who were in charge to comply with the requirements of the patent office. Their accusations that anything has been held back is foolishness and can be so proved by written evidence in my possession. The Patent Office has not made such a complaint. I assure everyone that anyone who makes such statements are not telling the truth and that it has been and shall continue to be my aim to see to it that there shall be no delay that I can help in getting patents on my inventions. The only delays so far have been caused by the lack of finances over which I have had no control. Finances are at this time the cause of delays. We can move no faster than we have means to do so. Reason and good judgment must be used and we cannot rush blindly on not knowing where we are going nor how we are going to get there because of the lack of finances. I would like to make a part of this record the recent letter from the patent office and my answer to the same. This does not sound like the patent offices is claiming that any data has been withheld. It was stated at the trial in Reno that others had tried to make the device and it would not work. It was known before starting that they did not have the material for the valve and therefore they could not make it work and that there were other very positive reasons why it would not work but which were theirs, not my fault. I have written evidence to prove this statement. However, that does not prove that all had not been disclosed. I shall at the first chance I get go to Washington, D. C., and see for myself just what has been done there. I have been kept in ignorance both as to the patent situation and the condition of the Moray Products Company affairs.

At this time we are not in possession of the Corporation books, but we are going ahead and will take the proper legal steps to get possession of them. I call on all the officers and investors to get behind me and I promise that I will do all in my power to put my inventions over in a successful way both for the investor and the public. From now on we want facts not untruths, cooperation not friction, all working for the one purpose. I have contracts with the Moray Products Company. In these I have been blocked at every turn and unreasonable requests have been made of me that it was clearly understood from the first I would never consent to as I have always felt they would mean the loss of everything. I have more than lived up to my contracts. I will continue to do so if conditions have not been created that make success under them impossible. I claim these contracts have been violated by those who have been in office to the extent that they almost if they did not fully accomplish their aim to either run or ruin. Stock has been thrown right and left at from a few cents to giving it away until it is hard to say just what will be necessary to protect the honest investors' money.
I am not discouraged, never felt better, and I am convinced of the final success of my efforts, and assure all that if and when I make and realize on my efforts that the investors shall also realize their full proto share."

It was the unanimous decision of the directors present to remove by this action all officers and directors, and officials of whatsoever nature and kind other than those stated above, particularly offices that may have been held by Messrs. D. V. Farnsworth, C. Fred Schade, E. H. Hursh, H. C. Carlisle, H. B. Carlisle, Murray O. Hayes and any others that may have been appointed to or acted as official of the Moray Products Company.

With this action of removal from office there is further the definite declaration made that the present and only officers of the Moray Products Company are: T. Henry Moray, Director and President; Walter Rowson, Director and Vice-President; S. E. Bringhurst, Director and Secretary and Treasurer.

As a matter of record and a part of these minutes the Court's Decision signed at Reno, Nevada, July 22, 1932, is herein written in detail.

No. 39886 Dept. No. 2.
IN THE SECOND JUDICIAL DISTRICT COURT OF THE STATE OF NEVADA, IN AND FOR THE COUNTY OF WASHOE

IN RE MORAY PRODUCTS COMPANY, a corporation.

Amended Judgment and Decree

This matter coming on regularly to be heard on the 25th day of May, 1932, upon the petition of Thomas Henry Moray, said petitioner appearing by Messrs. Brown & Belford, his attorneys, and Moray Products Company, a corporation, together with certain stockholders therein, appearing by Messrs. Walter Rowson and Ernest S. Brown, their attorneys, and oral and documentary evidence having been introduced and the matter having been argued to the Court by counsel for the respective parties, and the Court having duly considered the same, and it appearing to the Court that due and legal notice of the hearing of said petition has been given, as required by that certain order duly made and entered in the above entitled court and cause on the 23rd of April, 1932, and it further appearing to the Court that said Moray Products Company has failed to elect directors within six months after the time designated for its annual meeting. to-wit, the third Monday of October, 1931, **and it further appearing to the Court that the petitioner, Thomas Henry Moray, is a stockholder of said corporation and holds stock entitling him to exercise a majority of the voting power of said corporation,** and it further appearing to the Court that the principal office of said corporation is located in the City of Reno, County of Washoe, State of Nevada;

IT IS HEREBY ORDERED, ADJUDGED AND DECREED: That E. H. Hursh, C. Fred Schade and Murray O. Hayes be, and they hereby are removed as directors of said Moray Products Company, a corporation.

IT IS FURTHER ORDERED, ADJUDGED AND DECREED: That Thomas Henry Moray. S. E. Bringhurst and Walter Rowson be, and they hereby are appointed directors of said Moray Products Company, a corporation, and tha' the Board of Directors of said corporatoin shall consist of said directors, and that said appointees shall have the same rights, powers and duties, and the same tenure of office as directors duly elected by the stockholders at the annual meeting held at the time prescribed therefor-to-wit, the third Monday of October, 1931, would have had.

IT IS FURTHER ORDERED, ADJUDGED AND DECREED: That petitioner shall have and recover his costs of suit.

DATED THIS 22nd day of July, 1932.
(Signed) D. F. CURLER,
District Judge.

It was unanimously decided that the working office of the Moray Products Company at which place files, records, and accounts are to be kept will be located at Salt Lake City, Utah, however, with due consideration to complying with all the requirements of the State of Nevada under which the Moray Products Company was incorporated.

Consideration was given to the suggestion of the appointment of additional directors, and it was decided that for the present and until the books, records and affairs of the old officers could be audited and presented to the Company, we should recognize only the directors appointed by the Court's Decision.

* Demand having been made on Director Rowson for the old records, certificates, minutes and files of the company, and he having deneid the request and continues in holding the same, it was decided to order New Certificates printed,—new records started, and to work out a plan to protect all having legitimately secured certificates by issuing new certificates therefor; and it is declared that the Moray Products Company cannot and must not be held accountable for certificates unless legitimately issued.

Motion was made and authority given to have all books and records audited when secured.

It was decided at the next directors' meeting a resolution would be adopted amending the By-laws to read in:

Article 1, Section 6: Special meeting of the stockholders may be held when called by the President or by a majority of the directors of the corporation or by a majority of the issued and outstanding stock of the corporation.
Meeting adjourned sine die.

(Signed) S. E. BRINGHURST,
Secretary.

No................. Dept. No.................

IN THE SECOND JUDICIAL DISTRICT COURT OF THE STATE OF NEVADA, IN AND FOR THE COUNTY OF WASHOE IN RE MORAY PRODUCTS COMPANY, a corporation

Judgment and Decree

This matter coming on regularly to be heard the 25th day of May, 1932, upon the petition of Thomas Henry Moray, said petitioner appearing by Messrs. Brown & Belford, his attorneys, and Moray Products Company, a corporation, together with certain stockholders therein, appearing by Messrs. Walter Rowson and Ernest S. Brown, their attorneys, and oral and documentary evidence having been introduced, and the matter having been argued to the Court by counsel for the respective parties, and the Court having duly considered the same, and it appearing to the Court that due and legal notice of the hearing of said petition has been given, as required by that certain order duly made and entered in the above entitled court and cause on the 23rd day of April, 1932, and it further appearing to the Court that said Moray Products Company has failed to elect directors within six months after the time designated for its annual meeting, to-wit, the third Monday of October, 1931, and it further appearing to the Court that the petitioner, Thomas Henry Moray, is a stockholder of said corporation and holds stock entitling him to exercise a majority of the voting power of said corporation, and it further appearing to the Court that the principal office of said corporation is located in the City of Reno, County of Washoe, State of Nevada;

IT IS HEREBY ORDERED, ADJUDGED AND DECREED; That D. V. Farnsworth, C. Fred Schade and Murray O. Hayes be, and they hereby are removed as directors of said Moray Products Company, a corporation.

IT IS FURTHER ORDERED, ADJUDGED AND DECREED: That Thomas Henry Moray, S. E. Bringhurst and Walter Rowson be, and they hereby are appointed directors of said Moray Products Company, a corporation, and that said appointees shall have the same rights, powers and duties, and the same tenure of office as directors duly elected by the stockholders at the annual meeting held at the time prescribed therefor, to-wit, the third Monday of October, 1931, would have had.

IT IS FURTHER ORDERED, ADJUDGED AND DECREED: That petitioner shall have and recover his costs of suit.

DATED this 11th day of June, 1932.
BEN F. CURLER,
District Judge.

ASSIGNMENT

FOR AND IN CONSIDERATION of one million and fifty thousand (1,050,000) shares of the capital stock of the Moray

Products Company, a corporation of Nevada, to be issued and delivered to me or my order by said corporation, and other valuable consideration, receipt of which I hereby acknowledge, I hereby give, grant, transfer, sell, assign and set over to said Moray Products Company, all of my rights, title, interest and equity in and to the following described property: Any and all inventions and improvements thereto, processes and discoveries, chemical and/or mechanical combinations, and formulas, now or heretofore belonging to the undersigned T. Henry (T. H.) Moray; and any and all inventions, and improvements thereto, processes, discoveries, formulas, chemical and/or mechanical combinations secured under Letters Patent, trademarks, copy rights and rights in the nature thereof of the United States or of any other countries by the said T. Henry (T. H.) Moray, or any for which letters patent, trademarks, copy rights, and rights in the nature thereof, of the United States or other countries, applications have been, are now, or shall hereafter be made by or allowed to the said T. Henry (T. H.) Moray, together with any or all of said letters patent, trade-marks, copy rights or rights in the nature thereof of the United States or other countries; all completed machines of the said T. Henry (T. H.) Moray; and I hereby further agree to give to said corporation until October 1, 1940, my exclusive time, talents and ability in perfecting and completing said discoveries, chemical and/or mechanical combinations and improvements thereto and in making further scientific and other discoveries for such salary as is provided by the board of directors; and I further agree to give, transfer and assign to the said corporation all inventions, improvements thereto, discoveries, processes, formulas, chemical and/or mechanical combinations, patent rights, trade-marks, copy rights and rights in the nature thereof, which I may acquire any right, title, interest or equity in during a period of ten years following the date of incorporation of said company, all of which said property, aforementioned, is, however, transferred, assigned and sold to said corporation, and by it accepted, subject to one certain written contract dated the 1st day of October, 1930, copy of which is hereto attached, and made a part thereof, marked "Exhibit A," by and between T. H. Moray, party of the first part, and D. V. Farnsworth, C. Fred Schade, and Murray O. Hayes, parties of the second part, granting to parties of the second part the exclusive right, under the terms and conditions set forth in said agreement, to manufacture, sell and distribute all of the machines, discoveries, formulas, processes, chemical and/or mechanical combinations so owned or acquired by the said corporation, by paying to said corporation 10% royalty thereon.

It is further understood that the undersigned, T. Henry (T. H.) Moray shall retain personal ownership and exclusive charge and control of the laboratories and equipment used by him and his assistants for experimental and development purposes.

Signed, sealed and executed at Salt Lake City, Utah, this 31st day of January, 1931.

T. H. MORAY.

Witnesses:
Robert A. Stelter
D. Thomsen

STATE OF UTAH
COUNTY OF SALT LAKE
ss.

On this 31st day of January, 1931, personally appeared T. Henry (T. H.) Moray, the signed of the foregoing instrument, who duly acknowledged to me that he executed the same.

ROBERT A. STELTER,
Notary Public, Residing at
Salt Lake City, Utah.

My Com. exp. 3-6-1934.
(SEAL)

MURRAY O. HAYES
Lawyer
Salt Lake City, Utah

Jan. 10, 1931.

T. H. Moray,
2484 So. Fifth East.
City.

Dear Henry:

We the undersigned, D. V. Farnsworth, C. Fred Schade, and Murray O. Hayes, do hereby promise and agree that, as directors of the Moray Products Company, we will vote to have the said company pay you a minimum salary of $1000.00 per month as soon as the company can do so, consistently with due regard for the safety and stability of the company, **and from the time of the organization thereof until it is able to pay the aforesaid salary, to pay you sufficient sums to meet your indebtedness and provide you with sufficient money for keeping yourself and family.**

D. V. FARNSWORTH,
C. FRED SCHADE,
MURRAY O. HAYES.

SALES AND MANUFACTURING CONTRACT

There were two contracts (the one expired January 1, 1932, which gave them the right to sell stock under certain conditions) that were prepared by Farnsworth, Schade and Hayes.

One was for the sale of certain block

or blocks of stock in which it was agreed that Farnsworth and Schade were to sell sufficient stock to net me or the corporation the sum of $10,000.00 on or before January 1, 1932. To guarantee this sale Farnsworth and his wife executed a deed to certain ranch property in Mexico in my favor. They placed the value of this ranch at $15,000.00. This deed was held in escrow by M. O. Hayes. (You see, technically, the deed was never delivered to me, as M. O. Hayes was a party to the contract and was one of the second parties. Furthermore, it was impossible under the laws of Mexico for them to have deeded this property to me.)

I told them I had no interest in this ranch, that they were doing this thing without my request or desire and for their own purposes, and that if they should fail to sell the required amount of stock it would be impossible for me to get the ranch anyway.

M. O. Hayes and I, to please Farnsworth and to quiet the fears of his wife and his friends that he would lose this ranch in Mexico, released him from the jeopardy (which he called it) of losing this ranch, as he claimed to have fulfilled that part of the agreement as to the raising of the $10,000.00 in the sale of stock.

(This ranch transaction being wholly a matter connected with their guarantee in the sale of stock and nothing to do with the so called "Sales and Mfg." contract, an illegal manipulation of the requirements necessary for a good contract.)

I want to point out here that this $15,000.00 claimed, and it is so stated in the contract, consisted of the execution of the above mentioned deed and that no claim is made, nor was any cash or any consideration received for any contract.

Both of these contract Hayes brought to my home when I was sick, and he asked me to sign them. I told him I was not well enough to read them, and he said, "Henry, can you not trust me? And if at any time you decide you wish the sales and manufacturing contract back, I will sell it to you for $1.00. I can do this, you know, because the actions of one partner are binding on the others, and this sales and manufacturing contract is strictly a partnership arrangement between the three of us, i. e., Farnsworth, Schade and Hayes, and I am always in a position to protect you, being in this partnership arrangement."

As time went on I saw that Schade and Farnsworth were using this contract for purposes other than that agreed upon; that they were not using it to protect the company, as they had first represented to me that it was necessary to have this contract for the protection of the company, but that they were attempting to capitalize on it personally.

I took this matter up with Hayes, pointed out where they had violated their contract, and he agreed with me and I said I would revoke the contract.

With this thought in mind, on April 21, 1931, I wrote Hayes a certain letter, copy of which I am enclosing and which I will refer to as letter "A." Hayes requested that I leave the matter in his hands, again asking me to trust him, and pointing out that the actions of one partner are binding upon the others and that inasmuch as I could trust him I had nothing to fear as he would always be there to protect me and to give back the sales and manufacturing rights, and that anyway there had been no consideration paid me for the contract. (It must be remembered that Hayes had passed the bar examination and as an attorney he was supposed to know that the advice he was giving me was good and legal.)

After Hayes went to Washington I wrote and told him the conditions were getting so bad that they simply could not go on between the other parties and myself and I asked him to return to me, according to his promise, the sales and manufacturing rights. I have a letter from Hayes, dated August 7, 1931, in which, among other things, he states, "I talked to the man who draws the contracts for the Navy Department, about the form of assignment for transferring the sales rights to you, and he said it cannot be done. He says that a contract is not partnership property and that the persons who are the parties of the one part to the contract are not partners, but merely engaged in a joint enterprise. I had no idea that contracts stood in this special status. If it has reached the point where you simply cannot get on with Schade, why do you not ask him to resign from the Company? If he will not, you have the control and can call a stockholder's meeting and vote him out, then he will have nothing to say about the Company business.

"I think Vernon Romney or Carlisle would be a good man, maybe it would be best to have both."

I would like to point out the second paragraph just quoted in Hayes' letter, and those acquainted with the recent law suits can readily see that I could not do the things mentioned in that paragraph and it is reasonable to believe that Hayes deliberately deceived me, as he was one of those who had to do with the incorporating of the company and the drawing up of the bylaws. In order for me to accomplish what it was my understanding could be easily done and which Hayes herein proves that he right along had given me to understand, yet I was unable to do this thing so easy, and it was necessary for me at great expense to go to court to accomplish something which I had been given to understand I could do by a simple act of my own.

Now to return to my request for the return of the sales and manufacturing contract.

You can see by the first paragraph just quoted above that Hayes has to acknowledge that he had grossly misinformed me as to his ability to return to me this contract. He later wrote me a letter telling me he was willing to return me this contract if I were willing to guarantee to support his family while he was in jail if Schade and Farnsworth should put him there for selling this contract to me, and many other ridiculous things he wrote in this letter.

After much correspondence back and forth, on February 6, 1932, in answer to a request from me that he at least return to me all right and title he had to this contract, he wrote me among other things the following:

> "What would it avail if I turned back my interests in the sales contract, with Farnsworth and Schade still holding theirs? Have they agreed to relinquish it? Why shoud I, who worked with you for four years before they heard about it, give it up and let them continue to benefit by it? As it is, no one can benefit by it unless you do also."

So I have been unable to get Hayes to in any way live up to what he promised. In the above paragraph it plainly shows he will not do what he has promised and could it be because of his greed—his fear someone else is going to get something that he may not, keeps him from doing the right thing Could it be he holds with those he knows are in the wrong because he deceived himself into believing they had me in a box? Perhaps they tell him they have me tied and he is in fear that he will lose if he does not hang with them. Does he sell himself thinking he will make millions if he hangs with those he thinks have sprung a trap on me and have me secure even when he knows they are wrong?

As stated above, the whole purpose of this sales and manufacturing contract was not intended for their individual, personal benefit, as they had given me nothing in return for this contract, but it was because Hayes and Schade and Farnsworth made the claim that it was the only way I could keep the power interests from tying up the invention.

Hayes was an attorney. I trusted him. When I told the three of them that I could not see where this Sales and Manufacturing contract offered any protection, they told me that Attorney Bowen and Attorney Clark said it was the only way, and Hayes, as an attorney, also said it was the only way.

I want to point out that in these agreements that very plainly state that the $15,000.00 was not cash, but the deed to this property, executed by Farnsworth and his wife, held by Hayes and given as a guarantee that $10,000.00 worth of stock would be sold on or before January 1, 1932, and that they make this same deed to this Mexican property the consideration for the Sales and Manufacturing contract, that it was not possible that through any violation of the Sales and Manufacturing contract for me ever to receive this Mexican property only through the violation or failure on their part to sell $10,000.00 worth of stock could I have received this property. The laws of Mexico make even that impossible. Again I want to point out that even if they had failed in the one agreement of the sale of stock, which was an absolute, separate and distinctly another agreement from the Sales and Manufacturing contract, I could never have received the property in Mexico anyway.

Then in the Sales and Manufacturing contract they turn around and make the claim that they have paid me $15,000.00 for the Sales and Manufacturing contract, but acknowledge that this $15,000.00 consideration is this same deed to the Mexican property mentioned in the other agreement, and no provision is made in this contract whereby I ever could have obtained this ranch. How in the name of reason can such an arrangement be considered a consideration for any contract?

They claim to have raised $10,000.00 in the sale of stock, and, which as pay for so doing they received 33⅓%, and that in fulfillment of that claim Murray Hayes, with my permission, returned the deed to the Mexican Ranch, which was a guarantee that they would sell $10,000.00 worth of stock on or before January 1, 1932. What money they did turn over to me was not theirs but received from the sale of my personal stock on which they received large commissions.

$10,000.00 is not $15,000.00, and I never received any consideration of any kind for the Sales and Manufacturing contract or any other contract.

Besides, they have failed to do a score of things they promised to do in other contracts, including the Sales and Manufacturing contract.

Can the Mexican ranch property be considered a consideration for any contract? Can it be considered a consideration for a guarantee of the sales of $10,000.00 worth of stock in the stock selling contract and at the same time be a consideration for a separate and another contract known as the Sales and Manufacturing contract not the sales of stock but the sale of the inventions after they are perfected? No provision being made on the second contract whereby I ever could have gotten the consideration.

August 27, 1932.

Mr. Walter Rowson, Attorney,
223 E. C. Lyon Building,
Reno, Nevada.
Dear Mr. Rowson:

I am writing you for two reasons.

First, as President of the Moray Products Company, and in accordance with the resolution passed at the directors meeting of July 29, 1932, I am requesting you to send to me at 2484 South 5th East, Salt Lake City, Utah, all books and records and papers, in fact everything in your possession that belongs to the Moray Products Company. This does not include, of course, such records as are required by law to remain in Nevada with the company agent, Mr. Brown. May I look for these records at once? Send them express, charges collect.

The second reason I am writing you is that I would like to be able to decide just how honest and sincere your convictions are in regard to the "controversial matter" as you call it, and just what this matter is.

I am giving you the benefit of every doubt, for I respect every man's honest opinion even when I know he is wrong. If I were fully convinced you were deceitful and of the same type as those I once trusted I would not try to work with you, for such would be useless, and I would request that you resign.

I must have those around me who know and understand the truth and will help, not hinder. Your attitude is dead wrong in regard to the true situation. You have based your conclusions and opposition wholly on the reports and statements of Farnsworth and the others who have not worried or thought it necessary to give their story the slightest resemblance of truth.

Your attitude would have to change if you expect to remain with me; for so long as you believe or support the same viewpoint as those removed you cannot help.

At no time have I done anything, nor will I do anything to hinder the progress of my inventions.

I am the one most interested and the one who will receive the greatest benefit by my inventions going ahead as fast as they can. The patent office has not made a request of me that I have not granted. I intend to grant every request made by the government that I can.

If you are honest in your belief, I would be willing to go all the way to prove to you that you have been deceived by untruths and misrepresentations.

I have never in all my life met with men who value the truth as little as Farnsworth, Schade, Carlisle and Hayes.

For your information, I would like to let you know that there has not been one cent of value invested in the Sales and Manufacturing Contract.

The parties have drawn large commissions for every effort they set forth to sell stock, and I have again informed Schade Farnsworth and Hayes that I do not and will not recognize them in any contract or in anything they do.

I have not changed my idea nor my policy on the patent or any other aim which was the basis on which I permitted the Moray Products Company to be formed.

I am in the right, am doing all I can to push things to a successful conclusion for every investor.

The falsehood of certain parties have not and shall not prevent me from going ahead to the fulfillment of the thing I set out to do.

I welcome suggestions, but will take commands from no man, and more especially when I know they come from the type of men who have only their own selfish ends and purposes to accomplish.

My policy has not changed and shall not so long as I believe myself in the right.

I am going ahead to success. I went into this for a very definite purpose, have let that purpose be known from the first, and see no reason why I should change.

Mr. Rowson, I have been frank because the time has come for united effort. You have fought me long enough. It is a matter of either you work with me and help me make a success of these inventions, as an officer of the Moray Products Company, or there will be a parting. I want only justice and right.

I want you to have a chance, for your benefit, not mine, to cleanse your mind of the false statements of others. With you it may be just another page in life. With me, this work is my life and I can permit nothing to stand in the way of success and the final and lasting benefit to mankind that my inventions will be.

You surely must realize that Moray is

the one man that cannot be replaced in an organization to work out the final success of the Moray inventions. There is not room for men who share the views and believe the untruths that certain ones have been circulating for their own selfish motives.

Hence, this letter stating that a thorough enlightenment of the truth will be necessary on your part if you are to be of use in the organization.

THM/T (Signed) T. H. MORAY.

August 30, 1932.

Mr. Walter Rowson,
Attorney and Counsellor-at-Law,
223-225 E. C. Lyon Building,
Reno, Nevada.

Dear Mr. Rowson:

Acknowledging receipt of your kind letter of August 26th.

By this time you have no doubt received my letter of August 27th, in which I endeavored, in as kind a way as possible, to place before you my attitude in the premises. I hope you will read carefully and conclude consistently what is necessary in regard to your relations with the Moray Products Company and with the undersigned, if there is a sincere desire to continue with the Company.

Now in answer to your kind communication. In the first place, let me state plainly that Messrs. Lillenquist and Penney were not my emissaries for conference with you. They are salesmen selling on commission only, selling my own personal stock until such a time as we can have the company in shape so that we can have regular and proper salesman selling company stock. They are selling my stock that I might have the necessary means to exist on and to continue my very earnest effort to continue with the patents and the perfecting of the same.

I am somewhat annoyed that Messrs. Lillenquist and Penney should have assumed the prerogative of being emissaries in regard to company matters of the Moray Products Company, as they know little of them and have no authority or right to discuss them or endeavor to present or determine what might be company policy matters.

In regard to visiting Reno. I have no particular reason for visiting Reno, — I have no aversion to visiting Reno,—and I know of no benefit that the company could secure by my expending money to visit Reno. Furthermore, and definitely, I have no money to pay out for such a visit.

I appreciate your letter, which you state is in the spirit of conciliation, but I must assure you that I am asking no concessions, and if you will read carefully my letter of August 27th you will note I am asking for cooperation, not conciliation and concessions, and I am asking for cooperation that we might continue definitely and properly as the Moray Products Company.

I am again asking that you definitely and positively please send the records of the Moray Products Company to Salt Lake City, as requested in my letter of August 27th, in order that we might have them properly audited and have the company placed right in its business affairs as quickly as possible.

Again referring to your kind letter. There needs to be nothing but constructive action,—there needs to be no controversy, —there needs to be no legal tanglements, —all that is asked for and all that is needed is cooperation.

Yours truly,

THM/T (Signed) T. H. MORAY.

P. S. Dear Mr. Rowson, if money had been available for me to make a trip to Reno, I would have been present at the trial, as it was far more important for me to be at Reno at that time than it is now or has been at any time.

Reno, Nevada,
June 1, 1932.

Dear Sir:

As incorporators of Moray Products Company we have at all times had a solicitous interest in the sales of the company's stock, whether made from the treasur or from Dr. Moray's personal holdings. From the outset, in our initial efforts to raise sufficient funds for the Company, to carry on, and to defray the expense of proceeding to patent the Company's Radiant Energy Device and other inventions, the response accorded us by yourself and other stockholders was most gratifying.

Had matters progressed as we had hoped, it was our expectation to have surprised the stockholders by personally repaying the full amount invested to every stockholder who had purchased stock at Five Dollars per share.

Unfortunately, due to the obstructionist policy followed by Dr. Moray and which has resulted in the litigation now in progress in the Nevada Courts, we have all been disappointed in our desire to see the Company definitely under way.

For that reason, and in lieu of the cash refund of your investment which we had hoped to make and tentatively arranged among ourselves, we have decided to donate to you from our personal stock holdings shares of stock in the amount now owned by you. In so doubling your stock holdings, without additional cost to you, we hope in some measure to off-set the unavoidable delay in placing your company on a paying basis.

Stock Certificate No................,issued in your name for................shares, is herewith enclosed.

Yours very truly,

(Signed) FARNSWORTH, SCHADE & HAYES.

Any stock issued under the above arrangement will never be recognized by me.

T. H. MORAY.

MURRAY O. HAYES
Lawyer
Salt Lake City, Utah

May 27, 1931.

W. H. Lovesy,
Newhouse Bldg.,
City.

Dear Mr. Lovesy:

This letter is written to supply to you some information you desire regarding the Moray radiant energy invention.

One question which is frequently raised is whether there is an available supply of radiant energy, and what is its source. The answer to this is to be found in the conclusions of some of our eminent scientists, among whom Tesla is perhaps as well known as any, and whose knowledge of high frequency is probably as extensive and thorough as that of any living man. In his book "Experiments With Alternate Currents of High Potential and High Frequency," published in 1904, he says on pages 57 and 58 that there is a vast quantity of available energy in space, and that it will not be long until man will hitch their machines to the very wheelwork of nature.

Further, Dr. Gunn, a civilian scientist of the Naval Research Laboratory, has proved that the earth is generating 200,000,000 amperes of electric current.

The aurora borealis is considered to be very definitely an electrical phenomenon produced by the passage of electric charges through the rarified gases of the higher atmosphere.

The conversion of matter to energy in the stars is now very generally accepted as demonstrated, and reasoning from what occurs in radioactive disintegration, during which energy waves are radiated we may conclude that energy waves of very high frequency are sent out from the stars, one of which is our sun.

Moray has not proved the source of the energy which he obtains from his device, and it is possible that it is from both space and the earth, as he uses an antenna and a ground wire, and the power ceases to flow if either is disconnected from the machine. As in the reception of radio waves, so in this case the circuit is tuned by the right arrangement of inductances and capacities to respond to the particular wave frequency which it is desired to intercept. He uses a valve to prevent return of the power to the outer circuit and force it to go through the power application circuit.

The new hook-up of inductances, capacities, and resistances, in association with the peculiar valve already mentioned and the unique and very remarkable oscillator tubes of his invention, now make available 4,000 watts of electric energy, and he asserts that given the requisite materials to modify his present device he can produce 50,000 watts from one machine.

All that is necessary to put the device in operation when properly assembled and connected to antenna and ground wire is to start electric oscillations in the circuit, which may be done in many ways, but one of the simplest is to pass a few sparks from an electrophorous disk into the circuit.

The frequency of the current is extremely high, as is shown by the brush discharge when either the antenna lead-in or the ground wire is disconnected from the machine. Certain difficulties of insulation inherent in high frequency currents are avoided by an ingenious device which reduced the frequency on the output side.

It is difficult to make a full presentation without stating some facts which it would be imprudent to reveal at the present time. However, some phenomena observed in connection with it may be mentioned.

The electric light globes (standard articles bought in the open market) become exceedingly hot when operated on this current, due to the fact that the gas in the globes becomes incandescent under the influence of the very high frequency, and because of the incandescent gas the luminosity is much higher than with ordinary current.

Ordinarily when a potential is applied to the terminals of a condenser a full charge is acquired almost instantaneously, but with the oscillators filling up somewhat as when water is poured into a bucket, the longer the voltage is applied the greater the charge taken, up to the maximum capacity for the given voltage.

The quantity of current passing through the secondary of the transformer is sufficient to burn up wire of that size if ordinary current were used, yet there is no heating of the secondary even though there is no circulation of air through it to cool it, as it is completely enclosed in friction tape. All parts of the machine run absolutely cool regardless of the length of time operated.

It makes no difference whether one fifty

watt lamp is used or whether fifty of them are connected to the machine, the draft of current is adapted to the load.

There is not a sound from the machine when in operation, there being no moving parts. The entire mechanism is enclosed in a box 7x7x27 inches, including the transformer.

It cannot be derived from batteries, as it can be transformed and only an alternating current can be so handled, and battery current is direct.

It cannot be induction from adjacent power lines, as an induced current is always of the same frequency as the inducing current, and there is no line on earth carrying current of the frequency of this. It is not radio power, being much too great; even to operate a loud speaker it is necessary to amplify a radio current many times, to say nothing of lighting a lamp or heating a flat iron, and further the frequency of this current is higher than radio currents.

Hoping this brief exposition will be of assistance to you, I am,

Very truly yours,
(Signed) MURRAY O. HAYES.

MURRAY O. HAYES
Lawyer
Salt Lake City, Utah
May 25, 1931.

W. H. Lovesy,
Newhouse Bldg.,
City.
Dear Mr. Lovesy:

This letter is in compliance with your request for some facts relating to the Moray radiant energy invention.

A short time ago a demonstration was given for a nationally known physicist, a professor in one of the greatest universities of the country, at which time he noted and mentioned the following.

That when the oscillators are connected in the circuit the condensers fill slowly, and the longer the current is applied to charge them the greater the charge they take, up to their maximum for the applied voltage, much as in filling a bucket by pouring the water into it, instead of taking the charge practically instantaneously as is ordinarily the case.

That the lights are much hotter than is ever the case with any current used at the present time, due to the incandescence of the gas in the globes under the influenece of the very high frequency.

That the size of wire in the transformer could not carry the amperage passing through it without burning up, if ordinary current were used, yet the wires remained absolutely cool no matter how long the machine operated.

The above points show that there is developed something which is entirely out of the ordinary, and can not be accounted for on the basis of induction from existing power lines or current from batteries.

I believe the above, in addition to the presentation made in the letter whereof I gave you a copy a few days ago will enable anyone to form a reliable opinion as to this device.

Very truly yours,
(Signed) MURRAY O. HAYES.

UTAH OIL REFINING COMPANY
Salt Lake City, Utah
June 29, 1932.

Mr. T. Henry Moray,
2484 So. 5th East.,
Salt Lake City, Utah.

Dear Henry:

I have handed you the original letter of Murray O. Hayes, dated October 24, 1929, in which he advised of his being shown and having explained to him the wiring diagram of the assembly, and also the fundamentals of the detector, and in which he stated plainly that while it would not have been noticed unless pointed out by you, at the same time he had become familiar with all of the detail of the construction of your device for utilizing cosmic energy, and added that you had applied only fundamental principles of electric circuits.

There is attached a second letter in which Mr. Hayes furnished me a detail of his educational record.

The memorandums on the face of these two original documents were made by me at the time I was consulting with Murray O. Hayes in my office and, at the time, he **definitely advised me that he had been shown every detail of your invention and added that he could make one of the machines himself from the information that had been given him.**

Murray O. Hayes made these similar statements to me several times, and it may be from my determination in having him repeatedly make the declaration that he had been shown every detail of the invention that would give him the impression that I had a doubt about the invention.

Personally, I have never had a doubt, but I thought the real detail secrets of the invention were held only by yourself and after my conference at New York with Harvey Fletcher and Carl Eyring, at the Bell Laboratories, — having agreed with them that I would endeavor to induce you to give the detail of your invention to Murray O. Hayes,—I did, as I have stated, repeatedly asked Murray O. Hayes if he could duplicate your machine, and in every instance his answer was emphatic and to the effect that he could do so.

Yours truly,
W. H. LOVESY.

August 17, 1932.

Your letter of August 4th was handed to me by Mr. Moray for answer, and this on account of the fact that I have been endeavoring to help him in the premises and relieve him of letter writing, etc.

If any reasonable explanation was made to you at the time the stock was sold you, it would start out with the explanation that the incorporated value of the stock was only a means to an end of avoiding a high State Tax on values that were to be developed later, and if you have purchased any amounts of stocks you would, of course, realize that the par price has nothing to do with the selling price.

It has been agreed by those who control and have interest in and have stock of the Moray Products Company that the stock shall not be sold for less than $5.00 per share.

Mr. Moray is applying every cent of money he receives for the sale of this stock,—which is his own personal stock,—to further the development and with a desire to perfect his invention.

I have spent a considerable amount of time with him and in his laboratory and with his correspondence, and have written most of the letters to the Patent Office and know positively whereof I speak when I say that things are progressing nicely and there is a desire and hope that they do continue to progress until a very favorable result is secured for those who have invested in the enterprise.

Unfortunately, for Mr. Moray, he got into the hands of promoters who wanted to sell stock and had only a desire to sell stock and were not interested in the development of his patent or perfecting of his machine. It has been a source of considerable expense and great annoyance. However, he has triumphed up to this time, and I feel very confident that he will continue to do so because in my several years connection with and knowledge of his efforts I have known him solely and wholly to be honest and earnest in his efforts.

Now if anybody could convince you or any other man who had money to invest that they would secure a million times their investment after a brief period of time, the stocks would sell like wild fire. However, the honest investor of stock should be satisfied if there is an honest and earnest effort made to accomplish a desire for a favorable end.

Hoping that I have explained this matter and helped in the premises, I am,

Yours truly,
(Signed) W. H. LOVESY.

WHL/T

Salt Lake City, Utah,
August 29, 1932.

To Whom It May Concern:

It has come to my attention that the claim is being made that T. H. Moray is reported to have made demonstrations in my "M" Men's Mutual class of his Radiant Energy Device.

This is to state that at no time has T. H. Moray been in my class in the Forest Dale Ward for any reason what-so-ever and none of his inventions have been shown or even talked of by him or anyone representing him.

I personally have never seen Mr. Moray's inventions and know nothing about them for the past fifteen years altho I have known Mr. Moray for over twenty years.

Very truly yours,
PARNELL HINCKLEY.

BROWN & BELFORD
Attorneys at Law
First National Bank Building
Reno, Nevada

July Eleventh,
Nineteen Thirty-two.

Burton W. Musser, Esq.,
C/o Messrs. Ball, Musser & Mitchell,
804 Newhouse Building,
Salt Lake City, Utah.

In re Moray Products Company

Dear Mr. Musser:

This will acknowledge receipt of your letter of July 9th, and also your letter of June 28th which Judge Brown answered. I have just gotten back from Chicago and went over to see Rowsen today.

As stated to you in my letter of June 9th, Rowsen informed me that a meeting of the directors had been held and that the number of directors had been increased to seven. Since this was done while these men were directors, I saw no way of attacking its validity. However, he today informed me that what he said was that they had been considering increasing the board. This is not at all my understanding of our conversation, and if the facts are as he states them I can see no reaosn why I should have agreed with him to permit them representation on the new board. As Judge Brown wrote you, E. H. Hursh has been appointed a director.

In view of our understanding, it appears to me that the only thing Rowsen can decently do is to resign, but this he refuses to do. He asked me to delay any action until he could take up the matter with Farnsworth, Schade and Hayes, two of whom are in the East. I informed him that I intend to institute proceedings immediately, and if he could arrange matters in the meantime all well and good. I am exceedingly chagrined that matters have taken the course which they have, and am determined to force the matter to as speedy a conclusion as possible.

I am preparing and will file a motion to modify the decree hereinbefore entered, removing Hursh and substituting one of Moray's nominees. The motion will be based on the ground of fraud and deceit, reciting the facts that Farnsworth, Schade, Hayes and Hursh at all times knew that there were four directors instead of three, and wilfully and fraudulently deceived the court. It is my personal opinion that Judge Curler will not look kindly upon these men playing ducks and drakes with his court.

I was sorry not to see you in Salt Lake, but was in a hurry both going and coming.

With kind personal regards, I remain

Very truly yours,

(Signed) JOHN S. BELFORD.

JSB/H

BROWN & BELFORD
Attorneys at Law
First National Bank Building
Reno, Nevada

July twenty-first,
Nineteen Thirty-two.

Burton W. Musser, Esq.,
C/o Messrs. Ball, Musser & Mitchell,
Attorneys at Law,
804 Newhouse Building,
Salt Lake City, Utah.

In re Moray Products Company

Dear Mr. Musser:

I today argued the motion to amend the judgment, and the Court granted our motion. I am preparing form of judgment and will send you a copy thereof as soon as the same is signed.

You will recall that I wrote you to the effect that we had agreed to permit the minority stockholders to have three members on the Board of Directors, inasmuch as it had been increased to seven. In view of Rowsen's disregard of our oral stipulation to this effect, I feel that I am not bound in any way to recognize it and have so informed Rowsen, consequently you can fill the vacancies of the Board or not, just as you see fit.

The whole thing was an example of shiftiness and bad faith, and you may be assured that none of it escaped Curler. I took particular pleasure in informing the Court in my argument as to how I regarded that sort of thing in the practice of law.

I am now making an effort to secure the books of the corporation from Rowsen, and will write you as to this within a few days.

With kind regards, I remain

Very truly yours,

(Signed) JOHN S. BELFORD.

JSB/H.

BROWN & BELFORD
Attorneys at Law
First National Bank Building
Reno, Nevada

July Twenty-fifth,
Nineteen Thirty-two.

Burton W. Musser, Esq.,
C/o Messrs. Ball, Musser & Mitchell,
Attorneys at Law,
804 Newhouse Building,
Salt Lake City, Utah.

Dear Mr. Musser:

I today argued the Motion to Strike the amended complaint in the action of Moray Products Company v. Moray. The motion was granted without leave to amend, so I presume that this phase of the matter is cleared up unless Rowson takes an appeal. **If he gets too troublesome about the matter all that is necessary, of course, is to have the directors order him to discontinue the action.**

With kind personal regards, I remain,

Very truly yours,

(Signed) JOHN S. BELFORD.

JSB/H.

The Moray Products Company, beset by double dealing, finally necessitated that Henry Moray take legal action. The reader should particularly note the court case and results at the end of "The Patent Situation." The amended decree was found in Henry Moray's favor, and all the ownership for the patent rights were returned to Moray from the Moray Products Company. The Moray Products Company lasted only a short while after this. Henry Moray was given complete control of the corporation and finally it was dissolved. The company was unable to function because of the bad publicity and background it had been given in its court involvements.

A number of other corporations were formed by Henry over the next few years, from 1932 to 1936. First, one known as the Moray Manufacturing Company or Manufacturing Company was formed and, secondly, an ElRay Corporation (the El came from Ella, my mother's name, and Ray came from the electromagnetic rays). A Cosmoray Corporation was also formed, later contracted to "Cosray" when Henry Moray found "Cosmoray" had already been registered as a trademark in Chicago. None of these corporations lasted any considerable length of time, and they played only a very minor role.

In 1936, Henry Moray formed the Research Institute. He intended to form a nonprofit corporation. The stock was gradually turned back to the Moray family by the original incorporators. None of the original incorporators paid for their stock but held it simply as a service to help Henry continue with his work.

Eventually resolving the maze of confusion that had resulted from failure of the Moray Products, all ownership and assignment was made to The Research Institute. At the present time The Research Institute has ownership of all Dr. Moray's discoveries.

My brother Richard and I would more than gladly accept help in developing Radiant Energy. However, we will not be forced into a compromising position because of any information people think they have.

Regardless of the fact that some of the corporations Dad was associated with did not succeed due to circumstances beyond his control, he did not intend that any of those individuals who genuinely tried to help him, or any of their heirs, should lose on the development of Radiant Energy. In the meantime, we have spent our lives to continue his efforts.

CHAPTER 5
PATENTS

One of the most backward of all our American laws is our patent law. The independent inventor is completely at the mercy of the large corporations, unable to protect himself from those who are able to gain a monopoly through their ability to amass large numbers of attorneys and huge amounts of paper. Patent law prevents the inventor from asserting his own interests in that he must first honor all the regulations and decisions of the examiners and, thereby, become subject to the bureaucracy that will eventually, I think, destroy the Republic. Our Congress has failed to interpret the intent of our Founding Fathers and has not implemented the law in the way those Founding Fathers intended. No one stops to ask why the Constitution mandates that we have a patent office. Consequently, we have developed without vision a series of "stumble-bum" procedures that leave us completely at the mercy of designing individuals.

In his early years Henry believed, because of the advice of his attorneys, that patents were essential for a young inventor, and that in order for him to be successful he had to take out these patents. Even though he attempted to patent a number of devices, he was unsuccessful in all his efforts except in the case of the electrotherapeutic device.

Part of the reason for inviting Carl Eyring and Harvey Fletcher to work on the Radiant Energy Device was to provide a way for Henry Moray to apply for patents. To assist him in this work, Harvey Fletcher introduced Henry Moray to Murray O. Hayes, Ph.D. and patent attorney. It is unreasonable to say that Carl Eyring and Harvey Fletcher did not receive complete disclosures when they were attempting to assist Henry in gaining a patent application. It is also unreasonable to say that Murray O. Hayes did not receive complete disclosure; as his patent attorney Hayes would have to know everything Henry knew. We also know that Hayes said to William Lovesy he (Hayes) would be able to reproduce the Radiant Energy Device. Consequently, from the patent standpoint, it becomes even more difficult for me to believe that Henry Moray did not consult with his advisors.

These applications ran into immediate difficulty not, as some assert, because Henry Moray did not make full disclosure, but because of the ignorance of the patent examiners themselves. The original patent application was first filed in 1931. This application was rejected on a number of bases. First (Figure 17), "Because no means was provided for causing the cathode to emit an appreciable number of electrons, the current produced in the cathode by the antenna will not heat the cathode to a temperature at which an appreciable number of electrons per second are emitted." (In other words, according to Thomas E. Robinson,

Commissioner of Patents, a solid state device, a transistor, cannot possibly work.) Second, because "No natural source of electric wave energy is known to the Examiner and proof of the existence of such a source is required." In other words, it was not enough for Moray to demonstrate the effect of the energy source; he also had to identify it, which he could not do.

It should be possible to hold a patent examiner and prosecuting attorney pecuniarily liable for their actions; then when the damage is done and the government cannot be held liable, the individuals themselves would be legally responsible for their prejudicial actions.

U. S. patent law is designed not to protect the inventor but primarily to protect the public from the monopoly eventuating when a discovery that is made is held by a few. Consequently, in appealing the examiner's decision, the inventor is put in such a position that he eventually has to demonstrate to the patent office. This demonstration in itself prevents the inventor from filing patent applications after one year (two years in Moray's time). If he files after the one year limit, he can lose his patent altogether.

The end result too often is that the inventor either applies prematurely for a patent, before he knows all the ramifications involved, or he does not get the rights due him because time restrictions prevent him from making further applications. Moreover, a United States patent only gives the holder a right to sue the infringer. If patent rights were designed truly to protect inventors, the government itself would prosecute for any infringement of the patent law.

Altogether, seven applications that directly involved the Radiant Energy Device were filed, all of which were rejected by the patent office on the claims that either no energy was available and that the applicant would have to prove a source of energy, or that Henry Moray did not provide means to heat his cold cathode devices.

In an effort to complete his patenting, Henry Moray prevailed upon some of his friends to furnish sufficient capital to enable him to go to Washington D.C. and consult a number of patent attorneys who, he hoped, could help him obtain patents. Dad thought this would be an excellent opportunity to take the entire family, and we traveled by automobile to Washington. The trip was not uneventful. Even though we drove a brand new car we burned out a main wheel bearing between Salt Lake and Denver, forcing us to spend the next three days in Denver until a bearing could be obtained and installed.

We spent long hours on the road to make up the time lost, putting us in Washington D.C. in the middle of the summer. The heat with the moisture affected our Western physical condition, so for a week we did not go out except in the evenings. But luck was with us. As far as sightseeing and learning of the Washington area, the trip could not have been more successful. By chance, we met a manager of one of the taxi companies. He enjoyed meeting us day after day at an appointed hour, taking the family around Washington. He opened many doors for us. We had an opportunity to get into the White House and other areas under conditions that the ordinary tourist was unable to do.

On the other hand, Henry Moray had more than a difficult time with patent attorneys as he spent hour after hour going from one to another. Finally he came to the offices of Lancaster, Allwine & Rommel. When he met Mr. Allwine he said, "I'd like to ask a question. Are all patent attorneys in cahoots? Why won't anybody listen to me?" Mr. Rommel listened to his story, listened to what he had done, listened to the number of demonstrations, and then he said, "The reason why no one wants to take your case, no one wants to be involved, is that you've already lost your patent rights." Then he explained to Henry Moray the meaning of Public Law in Section 4886 of the U.S. Patent Code. He explained to Henry that if demonstrations were truly tests, not demonstrations simply for convenience or pleasure or profit, and the inventor had improved his device after each of these tests, then they did not constitute public use. As a result, Henry Moray learned for the first time what a dangerous line he had been walking.

It was very fortunate that Henry Moray had kept copious notes showing everything he had done; showing that these demonstrations were not demonstrations, in fact, but tests wherein with each test he improved his device. After several days of discussion, Lancaster, Allwine & Rommel agreed to take the case.

In a letter to Henry Moray dated August 6, 1938, Mr. C. Ernest Allwine, stated:

"The material you left with me yesterday was so voluminous that I could do little more than review the material in a general way. The entire material, however, has been reviewed and while I do not feel qualified to comment on the soundness of your theories, the evidence is very convincing that you are in the process of developing a marvelous discovery irrespective of your own or anyone else's theory as to the manner in which it operates. Up to the time and prior to the time you actually have on file all U.S. and foreign applications which it is desired to file and sufficient technical data to complete patent applications to obtain full coverage, any demonstrations which are not experimental could endanger and cause you to lose all your patent rights to your invention (See Sec. 4886 U.S. Patent Law on Public Use).

"If the inventor uses his invention for profit and not by way of experiment that is public use (Elizabeth vs. Pavement Co., 97 U.S. 126 at 137, 1878: James vs. Bowen, 139 Fed. 556, 1905; Grenwalt vs. Stanley Co. of America, 54 Fed. 2nd 195 at p. 196, 1931) unless actual use resulting in profit is necessary to show the inventor how to perfect his invention, and unless he does perfect it in accordance with the teaching of such use (Sprague vs. Mfg. Co. 12 Fed. 721, 724, 1882). But experimental use becomes public use when it extends further either in time, or in number of instances, than is reasonably required to test the invention. (International Tooth Brush Co. vs. Gaylord, 140 U.S. 55, 63, 1891; A. Shraders Sons vs. Wein Salts Corp., 9 Fed. 2nd 306; Austin Machinery Co. vs. Buckeye Traction Ditcher Co., 13 Fed. 2nd 697; Wilke vs. Manhattan Rubber Mfg. 14 Fed. 2nd 52, 1937). Which holds that demonstration of several small pieces of tubing, without disclosing the process by which it is made, preliminary to execution of a contract, held not to amount to public use."

Allwine's opinion bore out some of Henry Moray's favorite testbook quotations:

"In regard to the question of what constitutes a public use of a manufacturing process or machine, it has been held that even though the public may not have access to the area of a factory where a machine or process is operated, the mere sale to the public of the products manufactured by the process or machine constitutes a 'public use' of such machine or process.

"In contrast, purely experimental use of an invention, to test its operability and suitability for the intended purpose, is not considered a 'public use' within the meaning of the law, even though the public may participate in the experimental trial of the invention. Whether any alleged experimental use by the public constitutes a bona fide experiment and not a 'public use or sale' such as may invalidate a patent is a question which in the last resort can only be resolved by a court reviewing all the evidence of the particular use. For this reason, it is far safer for the inventor to file his patent application before any use whatever has been made of his invention by the public.

"Presumably, under the ruling in Ex Parte Ordas, the patent office may consider classified use by the Government to be 'public use' under the statute, even though such use is officially considered 'Confidential' or 'Secret' because of a security order."*

"To constitute public use, only one specimen of the invention need have been publicly used, and only one person other than the inventor need know of that use. Furthermore, the use may be public even though the invention is not visible to the public, provided that this is the normal way of using it (a watch movement, for instance).

"In an analysis of Section 4886 it was stated that under certain conditions an invention having been in public use, or on sale, was a bar to patentability.

"These closely related phrases have been the subjects of much litigation and have required many court decisions to clarify their meanings. If the inventor allows his invention to be used by other persons generally, either with or without compensation, it will be in public use within the meaning of the statute. And that use is public which follows a transfer of the thing used from its inventor to the user, without reserving any control over it, or without expecting to make any change in it, or without any other restrictions."**

"If the inventor, or any other person who has not promised to keep the invention secret, uses the invention for profit, pleasure, or convenience, and not by way of experiment, that is a public use, unless resulting in profit is necessary to show the inventor how to perfect his invention, and unless he does perfect it in accordance with the teachings of such use, or makes a diligent effort to do so. But experimental use becomes public use when it extends further, either in time or in number of instances, than is reasonably

*__Ideas, Inventions and Patents,__ *Robert A. Buckles, 1957, John Wiley & Sons, Inc., New York*

**__Inventions and Their Management,__ *Berle and DeCamp, July 1937, International Testbook Co., Great Britain.*

required to test the invention. Nor will the fact that the inventor is but an employee in the place where he used his invention, or the fact that the profit goes primarily to his employer, prevent the operation of this rule."*

Over the years many people visited Dr. Moray's Laboratories with no other interest than to satisfy their curiosity as to whether the stories they had heard were true. From the time when he had his first laboratory in his home until after he moved into his new building on Fourth East, the number of curiosity seekers grew into the hundreds. He performed experiments with the Radiant Energy Device while onlookers watched to see it it really worked.

A few salient facts are evident in the patent business:

1. That under the law to give a demonstration in order to satisfy an individual as to whether a discovery is valid, is not sufficient justification to protect the inventor from losing his invention under Section 4886, U.S. Patent Codes.
2. That curiosity seekers request demonstrations to satisfy their own pleasure and convenience when they do not intend to assist in the development of the invention.
3. That there are those individuals who would, if they had the chance, sap another's brains and talents and even appropriate from him his very brainchild for their own use.

K.K. Steffenson, a local attorney, wrote Henry in 1939:

"Your request for a statement from me concerning certain matters pertaining to your invention will be very difficult to comply with, due primarily to the fact, that I am limited in library facilities. In order to even partially advise you, it would be necessary to make a thorough study of all the cases that pertain to your special problem. This is impossible for me to do as there is no library at my disposal that has these cases in it. However, I am suggesting certain principles that you must follow if you are to be assured that your patent rights will remain protected.

"Generally speaking, patent protection depends upon the facts and circumstances surrounding your particular case. There is no case that will be exactly on all fours with your case. All cases that can be cited will only be analogous to yours, but they will not be your case. Each patent case depends upon its special facts and circumstances. All a judge can do is to apply the general rules and regulations of the patent statute and controlling cases to the facts and circumstances presented by your evidence, and this is also by analogy and if the facts and circumstances are understood when presented to a court by the judge as you want them to be, your position will be sustained, but if he doesn't so understand them, then what. Until your particular case has been presented to a court of last resort, no man living can tell you just what a court will decide in your particular case. He can only make a guess and his guess will be just

Ibid.

about as right as the chances of getting heads when you toss a coin. All this does not mean that a good patent lawyer can't give you sound advice, based on decided cases, but such a lawyer or nobody else can predict, with any degree of certainty, the legal effects of all the facts and circumstances surrounding your case. Remember this, that until your case has been presented in a court of last resort and there fully decided there will always be present the big "IF".

"Excluding yourself, I feel that I am as well acquainted with the facts and circumstances of your case as anyone. I feel that "if" a court could see and understand the facts and circumstances surrounding your case as you and I see and understand them the law and his decision will and would sustain and protect your patent rights, but there is always this "IF".

"Now, Henry, you can't afford to take any or too many chances with this kind of an "IF". You will find plenty of friends, and I mean real friends, who may try honestly to diminish the size of this "IF". But remember all the argument in the world does not and cannot diminish the uncertainty of this "IF". It will never disappear until, as I have said, a court of last resort has decided your case. Your problem, as well as mine, is to be so careful and certain of our grounds that there can never be any catch, if ever you have to go in court.

"Now , if the foregoing contains good sound advice, and I think it does, just how far should I go in the citation of cases in support of any experiments of disclosure that you are asked to make to certain people, who want or even may have demonstrations and other disclosures, before they are willing to assist you financially? Here again you can be advised to advantage by a sound and honest patent lawyer. There is a large body of patent law that is pretty well established in this country which have interpreted the patent statutes on certain matters and if this advice be wisely followed you can do a lot of things that will not involve your patent safety. But to know just how far you may go with full safety is entirely another matter. It is the answer to this question of 'How far you can go?' that you desire advice from me.

"You may as well know it, to start with, that I will not pit my judgment as to what the patent law will permit you to do with safety, nor will I attempt to interpret the decided cases to the extent of saying, that you can do thus and so and that you can't do this and that. Your case involves far too many angles for me to even speculate as to what final form the law will take if you give demonstrations of any phase of it, if you ever come into court. Your patent procedure has not yet been definitely determined. You have an application pending concerning some phase of your invention, some parts of a completed invention are not in the patent office. Other phases of the invention are not in the patent office. Other phases of the invention are just in the experimental stage and other elements of the invention have not been fully established through standardization. Any one of these elements and possibly many others, as you proceed with standardization, may have to be changed here and modified there and so ad infinitum, before you will be able to tell with complete certainty, just what will be the final invention and the exact relationship between all the elements of your invention. Now I would not be at all smart, should I try to tell you just what you can do or can't do. In fact you don't know but what some proported 'experts', in whom the court has a great deal of confidence may try to make 'black' appear 'white' and that if you get too close to the line of public use or public disclosure, they may make a skillful attempt to pull you across this line, by interpretations of certain facts and circumstances that would influence the court sufficiently to decide your case against you. This patent safety depends entirely in staying far enough behind the line that such a situation could not possibly arise.

"Again, suppose that standardization should establish that your invention is fully completed now and you go on disclosing the invention by demonstrations and other means for more than two years from now, on the assumption that the disclosures are for experimental purposes. Could you get valid patent after two years, if some experts could establish a good presumption that you know that your invention was perfected. Of course, unless some other application for patent was in the patent office in the interim, this wouldn't make any difference. But how are you going to know what applications are in the patent office until your patent has been issued. Suppose again that you make disclosure sufficiently, in your demonstrations, to permit a clever party to get the idea behind your invention, what's going to stop him from revealing to the third party sufficient of this invention to enable that third party to file a patent so as to create an interference with matters (that you are not in the patent office as yet.) And then attempting to prove that you knew these matters for more than two years prior to your filing an application covering them. It's simple patent law that you couldn't get a patent if you lost the interference suit. Maybe this other party couldn't either. How can you foretell what evidence may be adduced at such a hearing?

"It is useless for me to attempt to enumerate all of the possible things that could occur, and until you are solidly in the patent office on your complete invention, with all its elements, phases, characteristics and parts, my advice to you is to be very careful as to how far you use demonstrations and disclosures in dealing with anybody.

"It is easy to cite cases that hold that experimental use is never a public use within the meaning of statute, but there is a question of fact as to what facts and circumstances would be construed experimental use and what as public use. These matters are questions of fact for some third party to determine, which offer patent lawyers the opportunity to exercise their skill in interpreting the question of what is experimental use and what constitutes public use. You see, Henry, there is never any certainty because all depends upon "IF". If you can establish beyond question an experimental use, of course the law is plain. But there always exists the right to interpret evidence. And the other side has a right to try and establish public use, with the same set of facts, if they do the law is plain. But there is always the "IF".

"Your case is further complicated by the fact that you want protection under foreign patent laws, who knows what the foreign patent law is and what foreign courts will do? We don't even know what the patent law in some foreign countries is, let alone what it will be tomorrow. If you want to protect your foreign rights you know that you can't make a disclosure of any kind that could be construed to be a publication in this country. Suppose for instance, that some party, to whom you make a demonstration and explain the operation of your invention and its main characteristics, and reveal some of the elements of the invention, such a party takes a notion to write up all of these disclosures and if this were published in a magazine or in a trade journal, and you never see it. Then what may happen? There would be an issue of fact raised, and that if you came into court, could be construed by some clever attorney to be a publication or a dedication to a public use. Then again we would have this monster "IF" again. As a matter of fact these things would never constitute a publication or a public dedication but we don't know that construction can be made by certain facts until after they have been tried out and determined by a court of last resort. It is these things that you have got to avoid, if you want to feel that you are fully protected. Your invention has any number of a hundred possibilities that do not depend upon the law, but depend upon how facts can be interpreted or misinterpreted by clever attorneys and how these interpretations will be construed by a court.

"Now, Henry I realize as fully as anyone that in order to get your invention financed you will possibly have to take certain chances in order to convince an investing type of mind or minds of the validity of your claims. And I realize that you want legal advice that you can rely on that will protect your invention from every possible angle. But I could cite you cases, that would possibly fill a volume, on what the law is but that isn't the point. The point is that constructions can be placed upon the facts and circumstances surrounding what you do so as to correctly fit into a case that is in your favor. I am fully convinced that I could cite you all the cases in the law that have been decided but they will all be based upon the facts and circumstances of some other invention and not of your case.

"In my humble judgment, I do not see much advantage in citing you a lot of cases. Your whole trouble, if you ever have any, is going to be with human nature, and not so much with what the patent law is or is not. You could be the cleverist lawyer in the world, and you will still have to contend with the interpretation of facts and human nature. You may know patent law from A to Z and you will still be confronted with the constructions to be placed upon the facts and circumstances of your particular case and the application by a court of the law applicable to these facts and circumstances. Nobody is able, before hand, to give you absolute certain and reliable advice on these matters.

"The only honest advice, and I think helpful advice that I can give you is that you get an application in the patent office covering as fully as you are prepared to go, every known phase and element of your invention protected by the strongest claims that can be presented at this time. And until this is done you're not safe either in this country or foreign countries.

"From the foregoing you can see how useless it would be for me to cite you a lot of cases on what constitutes 'public use', 'publication' and 'on sale' and all other matters involved. For after every case I cited you, I would have to attach a caution: 'Now IF what you have done or do or intend to do, can be construed not to be a public use, a publication, or on sale and so forth, you can do so, but you must be sure that no other construction of what you do or have done or intend to do, can possibly be given any other construction.' Or, it may be put this way, 'If you are positively certain that you can convince any court, in spite of all what anybody else may say or do, that your disclosures, through demonstrations and otherwise, were purely experimental use, then go ahead and do so and so.' Or, 'If the contracts you have, may have entered in and/or intend to enter into, the photographs sent out and so on, do not amount to putting your invention on sale, then go ahead and make contracts and publish photographs.' (Of course I feel that this hasn't been done as yet.) I could go through the whole category of the patent law in this way and cite cases but it would certainly be of no value or help to you.

"I realize, Henry, that this is not what you want but after studying your question for several days, I sincerely feel, that if I just tried to convince you or anyone else that the patent laws meant so and so, and cite cases substantiating the same, I would not only have misled you but deceived myself. I am thoroughly convinced that what I have said above is the most definite information that I can give you that will really be of any value. Now if you find the foregoing to be useless, you are at perfect liberty to treat it so, and if you still want cases I can furnish them to you within a reasonable length of time.

"I don't suppose there will be a patent lawyer or business man but what would laugh at the foregoing but nevertheless I am willing to stand firm on the advice contained in the foregoing and thoroughly believe that it is the only advice that you can follow with any degree of safety."

As a result of the advice received by his lawyers, Henry Moray introduced to Cosray Research Institute the policy of no demonstrations. He did not desire to become involved in litigation. He felt that the device had been tested sufficiently and that further public testing was impossible or at least dangerous, and would only end up in litigation. His stand was further confirmd by Sam E. Whitaker, Assistant Attorney General of the United States, in a letter he wrote February 7, 1939, at the request of an administrator of the REA, John M. Carmody. Furthermore, demonstrations became impossible after the last detector material burned out, although Henry Moray did not wish to emphasize this point. He felt more secure in arguing the public use. This could have been a psychological defense that he himself had erected because of his financial inability to re-build his device.

In an attempt to verify his position in the patent office, Henry Moray wrote a letter, April 10, 1954, requesting copies of his applications and asking for any subsequent applications that may have been filed on this subject. The patent office replied that the patent applications had been destroyed because the office was under the impression that the applications were abandoned. However, I have found since then that this information was incorrect, and from what information I have been able to ascertain, the truth is that these patent applications have all "disappeared" from the patent office.

From time to time Henry Moray obtained information from various people that patent examiners and clerks can have conficting interests. In fact, according to one certificate I have, the person claims that he was paid to keep large corporations informed of patent applications that would be of interest to that corporation.

Similar information was forwarded by Mr. John Y. Smith from San Francisco, California on April 17, 1931, to Murray O. Hayes in Salt Lake City. He spoke of a group of prospective investors that Mr. Smith hoped would contribute to the efforts of the Moray Products Company and said, "One of the parties interested formerly held a confidential position with the General Electric and later with the Westinghouse Co. He nearly took my breath when I told him regarding Moray's fears that the proposition might be stolen from the patent office. He said it will just as sure as you send it there, that the U.S. Patent office is 'honey combed' with employees of the General Electric, General Motors, and other large Co's. That he had helped steal valuable data from the patent office as the request of the above co's. He said you were crazy if you sent a description of this device to Washington before you had plenty of money to follow it through and influence enough to prevent a theft. So I confess after hearing him that I was in error scoffing at the fears of Moray."

It has been said that the first thing an inventor should do is apply for his patent and the last thing he wants is have it granted! Once granted, it becomes public, and for a small fee anyone can get a copy. An application for a patent, after all, only establishes date of discovery.

CHAPTER 6
GOVERNMENTAL SUPPORT

To understand Henry Moray's relationship with the Rural Electrification Administration and the United States Government, one should understand how he became allied with the U.S. Government in the first place. Political views of individuals around him meant very little in his early life. Although he was exposed to a number of radicals, he was unalarmed, not dreaming people could go to extreme measures to further their political ambitions. Only after becoming exposed to these radical views did he realize that some of his associates were truly not his friends, and that he was being used to further their political aims.

Early in his Radiant Energy work he was introduced to two men, John and Dan Magdiel, who once had openly admitted to being members of the Communist Party. Both of these men had participated in Communist Party organizations even before the Communist Party came to power in Russia. Later, when the political climate changed, John would deny his relationship to the party. In 1955, the last time I saw him, I was rather irritated, and in order to chide him I said, "John, you are a member of the Communist Party." He replied, "Oh no, not me!" When Henry applied for a secret clearance for the research laboratory in 1957, the relationship between Henry, the Magdiels, and the REA proved embarrassing. Henry's clearance was delayed until a security hearing was held. This hearing, not open to the public, cleared Henry of any Communist activity, and his secret clearance was granted.

In 1926, Dan went to Russia in an effort to interest the Russian government in Henry Moray's work. In 1929, a Russian, I believe a colonel in the Russian Army, Yakovlev, came to Salt Lake City to visit Dr. Moray. Yakovlev was given a complete demonstration of the Radiant Energy device in detail, making several tests as Dr. Moray directed in order to improve the device. He checked all the wires to insure that there was no internal source of electricity in the device and checked the device after it was put into operation. He made a statement that the device was high frequency and that it was producing at the time some 665 watts of energy. Quoting from Dr. Moray's notes we find the following:

> "He was permitted to remove the ground connection and antenna connection as many times as he wished, and the machine was permitted to die down. He then made tests to see that the energy was all gone. The machine was then again tuned and after the energy came back he was again permitted to make the same tests. After spending considerable time with the machine, the device was killed by hitting the table on which it stood with a hammer. This was done

at the suggestion of Mr. Yakovlev, as it had been explained to him just what would happen if this was done, and the results were as claimed they would be. About two and one half hours were taken up in the demonstration. After the demonstration, the machine was opened for Mr. Yakovlev to inspect; and his report will cover all of that even to the testing of the condensers and the holding in his hand the detector and tubes or oscillators, which he will tell you weighed not more than 8 ounces. All in all, he was given as complete a demonstration as it is possible to give without disclosing the secret. Several hours were spent the next day discussing the theory of the device with Mr. Yakovlev. Mr. Yakovlev left the next day for California and four days later returned to Salt Lake City, at which time he and I spent some hours going over my invention. He seemed very much impressed at what he saw."

Yakovlev then requested that Dr. Moray come to New York in November of 1929 to visit with his superiors at Amtorg Trading Mission. Interestingly enough, when Moray reached New York, they ended up at the offices of the General Electric Company in Schenectady, New York — always after working hours with no one else around.

At this point, Henry became alarmed. Why were they so secretive? Why were they at G.E.? Who were these men? What were their connections with Russia? Why did they insist he go to Russia? His questions remained unanswered. No actual agreement was ever reached. After completing their discussions it was determined that they would negotiate further at some future date. At this point, because of his suspicions and fears, and because of his alarm, he questioned these men's intentions. He was invited by Yakovlev to return to New York by sea, but because of his previous experience with sea sickness, he traveled by train and subsequently returned to Salt Lake City.

When Henry would tell his friends what happened, and even later as the 1938-40 period developed and he learned more of the political ambitions of these men, no one would believe him. Few conceived during that period that Communism was any threat. With apprehension, he tolerated the Magdiels. He allowed the contract to build the laboratory (Fig. 59) to go to them because of the insistence of his financial backers. It was not until years later that he became fully aware how completely radical their views were. They had an almost religious dedication to Communism and the glory of Russia: to them Mother Russia and the "heartland" were to be protected at all costs.

When Henry asked Yakovlev if he wanted to meet with Dan and John Magdiel, he answered, "Oh no, not me! They're too powerful for me."

In approximately 1953, about the time Beria was executed and Malenkov took control in Russia, John Magdiel wrote Henry from Mexico, where he was living in seclusion because of his political views, telling Moray that within the next "five days" great changes would take place in Russia, so that he, Magdiel, would have more influence. He said that it would be Henry's big chance for Radiant Energy — under Russian auspices, of course!

The enclosed letter (Figure 58) from Dan Magdiel reflects the radical views that so discouraged and alarmed Henry. I have used the letter even though it is

Fig. 57

Yakovlev's insignia that he gave to Dr. Moray

dated in 1951, as it is typical of some of the statements I remember hearing prior to that time.

Dan and John Magdiel persevered after the Yakovlev/New York incident, determined that something should be done about Moray's Radiant Energy. As the Roosevelt Administration began to have greater impact and the REA began to shift to a more liberal attitude, more contact occurred between Henry Moray and REA engineers. Ben W. Crim of the REA called Henry from Baker, Oregon, and asked for a demonstration. Moray agreed. When Crim arrived in Salt Lake City, Moray was very eager to give him information. He performed a number of tests, and Crim was very impressed with them. In January of 1938, he wrote Dr. Moray about the conference he and Henry had recently held and the notes Henry had allowed him to read. Mr. Crim made this statement to Henry: "I understand Mr. Magdiel and the administrator are going to have a conference on this subject this week."

At this point another REA engineer entered the picture, Frank P. Woods, who with J. J. Jurgensen of Greeley, Colorado, had originally seen Dr. Moray's research work in 1933 (Figure 53), and had thoroughly endorsed Dr. Moray's work at the time. Mr. Woods, who had become an engineer for the REA, renewed his acquaintance with Henry Moray.

Henry's memo gives his view of what transpired with the REA: "As early as February 1939, I began to fear, from personal contact, that many offices in

1

L.A. Dec. 29-51

Dr. Henry Moray D.S.C.
2484. SO. 5th East. St.
S. L. City. Utah.

Dear Henry!

I was just reading the paper and came acros this peace on atomic powr for generating Electrisity.

I am sure this clipings will interest you, acording to what we know they are a litle late in the day.

They hav had plenty of exsperimenta! mony a litle more than you hat.

You notis on one of the items I underskored with red, even if the had it ready to go and could get it free, it wauld upset the whole nation posibly start trowble.

You must not touch the Devils system, he will see to that.

Under seperat cover I am sending you the USSR buletin read page 688

Fig. 58

Letter from Dan Magdiel

2

that shoult Interest you— and to think that nation just emerged from the dark ages abouth 20 years ago when I was there, they wher just getting on their feet.

In 1941 Hitler said, " in 4 weeks maybe 6 weeks we will have them licking our boots— the millitary (in the world now hows) thought the same thing.

In the winter 41–42 the Russians drove Hittlers armys back 400 Kilometers before he coult dig in. So he swore I will get them, they can not humiliate me, But the Lord said to Ezekiel, phophesy aginst him— 38 Cap 4. Vers

I will turn thee back xxx if I have to poot hooks into thy jaws— in the 6 vers he is called Gamer (Germany) Togarmah—(Italy)— (the latin races) from Japhits son. but hitler was to be the gard or the leader now read the 8 vers— in latter years— in to a land ther is brought back from the sword (Revolution) aginst the mountins of Isreal. All the world think this means palastine, Israel have not ben in palastine for 2700 years, so the

Fig. 58

Letter from Dan Magdiel (continued)

3

~~Stop~~ ←

the Lord wanted to humiliate the 10 tribs so he used Hitler, But. he was allsogingto finich Hitler and he did. In the 11 vers note unwalled vilages, inpalastine they have walls. igenst those who are atrest— to take a spoil & pray— he could not find that in palastine. Now read the whole capter & the next capter very carfully severall tims and than read the 30 capter of Jeremia and Joel 2nd capter— in 11. vers note that the Lord allso has an army and it is very grate— "For he is strong tha executed the word of the Lord" who do you think that is?

Note how the Lord is speaking to his people & pleading with them note the 17 vers this is the cap. that Moroni told the phrophet Joseph was about to be fulfiled— he allso mentioned the 11. cap. of Esaiah read 1–6 vers than 10 vers ther is wher Lenine lais in glory. "and his rest shall be glorious." now read the next 11 & 12 vers, wher the Lord shall set his hand the 2nd tim to recover his the remnant of his people which shal be left from Asyria etc. now read the 113 sec. of D&C. and see what Lenine's job was, now finice with the 78 psalm

You have allready read the 14 cap of Esaiah = the Tsar of Rus

Russia & see wat the Lord was going to do to the poor Tsar,— 8–9 years ago Russia was not consid- red among the great nations— Now the world trembles— the greatest nation on eart and frantiely seeking for alies to help in pece— the Lord is doing things— D. going places. Happy new year— Dd

Fig. 58

Letter from Dan Magdiel (continued)

4.

I did not think I was going to
write much, but heres I have given
you a weeks job. well when you get
threw reading my referenses you will
know more than you do now.

I hope it will give you as
much of a thrill as it has me.
I saw it comming in 1917 and sins
than I have learned to acnowlidg the hand
of the Lord in all things, and if you are
on his side ther is not a thing he wont
let you in on, just ask him.

Well good look in 1952 tell all
the family halo and the best of
luckes for the future.

A frind of ours is coming to see you
Mrs Winnifred Thout from Ogden if she
cons do all you can for her. I have
great hope that you can save her.

Sinserly yours Daniel H. Magdiel

Fig. 58

Letter from Dan Magdiel (continued)

Washington had more radicals in them than I had the slightest idea of and it worried me. I expressed my fears to some of my associates in Salt Lake upon my return from Washington in February 1939 and later informed Dr.* Frazer whom the REA had called in as a scientific expert to consult with me on my work and as my bodyguard.

"In answer to my letter expressing my fears that I had gotten Radiant Energy mixed up with liberals and radicals which I could not go along with, Dr. Frazer wrote me on March 27, 1939, 'The only thing J.M. should not say or write is anything in connection with a threat. When he does that he is very wrong and should be told so, and in fact I have told him so upon many occasions. I realized that the so-called 'radicals' are feared out West, but the term is not feared here in the East except by a few 'reactionaries.' So it was I seemed to be surrounded with radicals, liberals, and what have you and their talk of 'their daring plans' and 'this man is one of us and that man is not.'

"It must be remembered I did not go to the REA, they came to me. After my return from Washington in February 1939, I wrote Mr. Franklin Wood that I did not care to be put on the REA payroll and I could not, nor could my Los Angeles associates, see the REA offer had much, if anything, to offer toward the perfecting of Radiant Energy.

"While I may not have fully agreed with U.S. Representative Thomas D. Winter (R) Kansas, when in December 1941, he publicly stated that the REA in Washington, D.C. had 'fallen into the hands of a gang of Communists, fellow-travelers and political second story workers who do not hesitate to sabotage the national defense program in the interest of preservation of their political theories,' for I had in 1939 met and become well acquainted with the REA administrator and deputy administrator and some others first in March 1939, and I must say I have never met two finer loyal Americans or gentlemen anywhere. There were those in the REA who did fit Congressman Winter's description. I do know there were those in Washington, D.C. more anxious to let Russia have RE than get it for America. They continually spoke of putting the President over the barrel and of their 'Daring Plan,' so in March 1940, I refused to continue my research under the REA proposals which included Russia."

On March 24, 1939, writing from 2000 Massachusetts Avenue, Washington, D.C., Woods stated, "I just received copy of your letter to the Administrator which was sent to me by Felix. I was deeply distressed at your conclusion—we do offer to help you get a patent, and who could do it better than the Government—and through the per diem expense ($25.00) we are going to pay you to develop it. This will not preclude your using it on any private commercial scale you may find it desirable to do."

On March 29, 1939, L.M. Smith wrote those in Washington who were trying to get Henry to accept the REA offer, "I was unable to urge him (Moray) to accept the proposition in its present form."

April 2, 1939, the REA telegraphed Henry: "Felix Frazer leaving by auto-

**Even though Frazer was introduced to Henry Moray as a "Dr." of Science, it turned out that he was not.*

mobile today for Salt Lake City. He will communicate with you en route as to arrival date. Suggest you await his arrival before making definite proposal."

Frazer, representing the REA, came to Salt Lake City and began investigating Henry and Radiant Energy. The following is copied from Henry Moray's notes made in April and May 1939: "In April, 1939, a department of the United States Government on its own initiative, sent a gentleman to Salt Lake City who had been introduced to me as Dr. Frazer. This gentleman was in the Moray laboratory for a total of about two months spending hours each day making tests and taking pictures of the Moray Radiant Energy discoveries, especially the Radiant Energy powered radio, and gathering information for written reports to his superiors in Washington, D.C. After about two weeks, he said it was necessary he go to Los Angeles, California, and it was necessary that I also go to California."

In California, further conferences were held. Henry returned to Utah in about one week, and 'Dr.' Frazer came back to Salt Lake in May of 1939, when he again continued tests and experiments. He was getting advice and suggestions on testing from two scientists at Columbia University in New York City. After days of these tests he said, "I have tested and tested until I am fully satisfied and feel there are no tests left to make. As long as I have the device I will want to run tests, and I wish you would destroy the damn thing so I cannot ask for any more."

Frazer returned to California and Henry again accompanied him. While in Calfornia, he mentioned the "Jensen" test where the RE current had been passed through a sheet of ¼" plate glass in series with the antenna, and even then the device operated with no hindrance to the flow of current. Frazer said if he could use the "glass" test, that would be all anyone in the Government could ever ask for. It was arranged that he should make such a test. The only requirement Henry made was the Government should furnish the glass for these tests so it could not be said he had "doctored" the glass. Twelve sheets of window glass were used. The current passed through this glass and the device operated as if the glass had not been in series in the circuit. Frazer said, "That is enough, we will never ask for another test." They returned to Salt Lake City. He gave Henry several copies of pictures he had taken of the Radiant Energy device being used to power the RE radio, which was capable of receiving radio programs over great distances—even from the antarctic where Admiral Byrd was broadcasting from Little America.

Then one morning in Henry's laboratory—totally unexpectedly—Frazer took a hammer and smashed the RE device. He said, "Now I cannot ask for any more tests!"

Here we have two men working quietly in the one room laboratory that served also as a machine shop, supposedly under the most friendly conditions, carrying out experiments on the device so that the REA could be assured that it had been thoroughly tested, when without warning Felix Frazer took advantage of the tools that were available to wield a hammer and maliciously destroy the RE device.

My own reaction would have been to shoot the man, and in my frustration I said so, but my father said, "What good would that have done? I had my gun and he had his. It would have been my word against his."

From what I remember of what I was told, Felix arrogantly walked out. Later when my father made protests, Felix claimed that Dad had destroyed his own device. Circumstances demanded that Dad continue to work with Felix in spite of what had happened.

The question is often asked why Felix Frazer would do this. What was his purpose? From the evidence, I am convinced that it was to force Henry into cooperating with him and his associates. He realized that Henry was not in a financial position to re-produce the device. He knew much of what went into the device and that the expense of reproducing its components was beyond Moray's financial capabilities and the capability of the facilities that were available to him.

Henry had refused to enter any contractual arrangements with the government up to this point. Felix's strategy in smashing the device may have been to move Henry into arrangements he had previously resisted.

Henry had shown confidence in Frazer by disclosing details of three of the Radiant Energy tubes; to an extent he permitted him to make detailed drawings of these tubes for patent purposes. He wrote then, "I think that speaks for itself as I have only permitted one other scientific man to ever get that far into details of the construction during my entire work with this research."*

At one time, Henry was asked by the REA to investigate the effects on human beings of high energy, high frequency current. (I use the term "high frequency" in its broadest sense.) Even then Henry was very much opposed to the inductive heating processes used by diathermy machines. He started his investigation by examining all the known electrotherapeutic devices being used in the 1939-40 period. In fact, he became disgusted at some of the devices he was required to investigate, because the "devices" bordered on psychic phenomena.

A number of reputable medical doctors and osteopaths in both Salt Lake City and Southern California cooperated with Henry in these investigations. As far as I am concerned, these physicians and osteopaths were some of the finest men I have ever met. They honestly believed in what they were trying to do to benefit mankind in the advancement of medical science.

A number of highly successful tests of an electrotherapeutic device that Henry had developed were run under the direction of these doctors. However, the results were not accepted by the medical profession, and Dr. Moray was criticized because he did not understand the "pecking order" of the profession.

I will not go into detail with regard to Henry Moray's electrotherapeutic device and his research with it. However, the reader is referred to the research by Robert O. Becker, M.D., of Syracuse, New York, published by the State University of New York Upstate Medical Center, the Veterans Administration Hospital, and Syracuse University. Suffice it to say that today research is verifying that without electrons available as furnished by ATP, and without the electrical reactions that take place in the body, none of the human functions could operate. Chemistry, after all, is founded on electricity, and the importance

**Harvey Fletcher*

of body electrical activity in its chemical processes is at last emerging.

As early as March 17, 1939, Henry wrote the REA: "Your proposal appears to accomplish little more than it gives me a few months of work in merely repeating what I have already done. Neither I nor my western associates are of the opinion the REA's offer will spell success." However, upon the promise of a laboratory built by private parties, he did say he would try working with the REA personnel. Other cash grants were to add up to $500.00 per month and eventually to not less than $250,000 for equipment. This promise was made in the presence of S.E. Bringhurst at various times. This help was never forthcoming because he did not become one of the "Brotherhood," "one with them," "one of the Daring Plan." The bills incurred in the building of the laboratory were unpaid and fell upon him when he could not see their "liberal" ideas.

In 1939, Henry wrote the following and sent it to Washington: "I wish I could have someone come out here from our Government who believes in the strength of our laws, who believes that the Government that Washington and Lincoln stood for is not going to perish from the earth; someone who believes in the continued strength and protection of our United States laws, including the patent laws, and with faith in the preservation of the principles and the Spirit of '76, someone who will not have a pet hobby of 'our Daring Plan' and who does not live for 'the expected change in our national structure' which 'will have taken place under the guidance of our honorary 33rd President, First Presidency of our New Regime.' For all these mad reasons they wish to control Radiant Energy to 'put the President over the barrel' and force their will upon the people."

One former high government official (Thomas R. Amlie, Congressman, Wisconsin) wrote Henry trying, as Henry said, "to convert me to doing what the REA wanted me to do—disregard all patent protection and patent laws so they could get everything away from me or to cause me to lose ownership in my discoveries so that I could not interest anyone in this work in order to make a proper success under our present way of Government... Two years ago I made the statement that democracy had probably less than five years to run in Europe and that the process of disintegration there was probably not more than five years further advanced than here."

In other words, he thought that America would, in five years or so, be controlled by the so-called "liberals." Therefore, Henry had no hope of his work succeeding, except in cooperation with them. He could put no confidence in the protection of the laws of the country as they had been.

One thing is certain, REA did not help Henry financially. Twenty-five dollars a day and a few loaned instruments are of little assistance in developing a major invention, particularly the greatest invention of all time and one with far-reaching effects on the entire world.

Dickering, floundering, hindering the inventor and running him into possible debt, bringing lawsuits and threatening him — are these activities "helpful" toward progress? It took Henry until 1949 to rid himself of the debt this episode cost him!

In 1942, a congressional investigation of the REA was held in Washington, D.C. by Congressman Charles A. Halleck of Indiana. The investigation was

Fig. 59

Dr. Moray's Laboratory built 1939-40
2505 S. 4th E. Salt Lake City, Utah 84115

looking into Communist subversive activity and how this was hindering the war effort. In his investigation, Halleck discovered Henry Moray's participation in the REA (Figure 60).

I specifically remember the evening this story came out. Dad had not seen the newspaper article when a newspaper reporter called him, read the article to him over the telephone, and asked him for comments. Dad was taken aback, as he believed that he himself should be invited to testify before the investigation committee, and he did not wish to speak publicly until he had had an opportunity to testify. Consequently, he refused to comment to the reporter. He was very disappointed because he never was called upon to testify to Halleck's committee. Rep. Halleck also made irresponsible comments about Henry Moray and Radiant Energy without any supporting knowledge of the facts.

As a result of the constant threat to his life, my father carried a gun with him at all times. He carried a .32 in his pocket, and whenever he walked from the house to the laboratory at night he would wear a 32/20 revolver. He was an excellent shot in the old "Western" sense, and I remember having seen him shoot from the hip and hit a crow at about 50 yards.

On three different occasions, he was attacked at his laboratory and shot his way out of the situation.

The incident of March 2, 1940, particularly stands out in my mind. Late that afternoon a friend of mine and I were playing on the front lawn of our home. My cousin was just starting up his car, which was parked beside my father's car in the garage, the two cars side by side; from the street one could not tell which car was being cranked or who was driving. Suddenly several men in a sedan turned in to the driveway and pulled guns as if they intended to fire on the car that was starting up in the garage. When my cousin backed out, the men could see that it was not my father, and they quickly drove away. I told my father about the incident and he laughed, trying to minimize it to prevent my worrying about it. He later took my two sisters and me to a movie at the Center Theatre and instructed us to call him when we came out of the movie so that he could come and pick us up. When the movie was finished, we called home and were told to wait there, we would be picked up by my mother. However, no one came, and we waited for several hours. Finally, my cousin Chester picked us up. When we arrived home we discovered that my father had been shot in the leg, and the doctor was there attending him. S.E. Bringhurst, the president of the company, was also there.

My father's account of what took place was very simple. He had gone to the laboratory just after dark and on leaving, had found that he had forgotten to pick up some material from the one office that was kept locked. Because the material was bulky, he left the front door open and went back to the office. As he fumbled with the keys in the dark, unlocking the office door, he had the impression that someone was coming up behind him. As he turned to see who it was, he was hit on the shoulder by a heavy object. Although the blow left his right arm numb and half-paralyzed, with his good left arm he grabbed his assailant by the head and held him pinned to his left side, entangling the assailant's gun in his overcoat. As the first man struggled, a second man

House Member Says REA Blocks National Defense

WASHINGTON, Dec. 1 (AP)—Representative Thomas D. Winter (R) Kansas, Monday demanded an investigation of the rural electrification administration, which he asserted was "obstructing national defense as surely as a paid saboteur."

"The true friends of farm electrification," he said in the house, "may as well realize, that this federal agency has fallen into the hands of a gang of communists, fellow travelers and political second-story workers who do not hesitate to sabotage the national defense program in the interests of preservation of their political theories and perpetuation of their pay roll."

Introducing a resolution calling for an investigation, Winter charged that the R E A, despite copper shortages which prevent installation of distribution lines in many rural areas, had "several million pounds" of copper stored in open fields in Texas. He said this copper was intended for transmission lines linking several rural cooperatives centering around Gilmer, Texas, with a generating plant the R E A hoped to build. Winter contended it would be at least two years before such a plant could be built.

He said R E A officials had ignored the necessity for copper "in their anxiety to set up a holding company system, supercooperatives by which the top holding company in Washington can perpetuate and expand its control over a large segment of the farming population."

Winter asserted nationalization of the power industry was the first objective of the socialists. He charged that the government seized the Grand river dam in Oklahoma after federal works "deliberately forced the Grand river dam authority into bankruptcy."

He said the executive order for this seizure was accomplished with creation of an Arkansas valley authority similar to T V A despite the failure of congress to approve a bill providing such an authority which was introduced nearly a year ago.

"This piece of legislation was simply leap-frogged over the congressional process and became a fact by dictatorial decree," the Kansan charged.

Winter said the administration had led this country into "a shooting war," but that it was the duty of all citizens, without regard to politics or partisan advantage, to support national foreign policies. He asserted, however, that "a large segment of the public mind sincerely doubts the good faith and integrity of the president and his administration," and added:

"National defense is being fantastically abused as the excuse for perpetuation and expansion of federal activities which divert from bonafide defense activities the energies and resources of the federal [illegible]ent."

Lawmaker Hits Bureau Tieup With S. L. Engineer

WASHINGTON, Dec. 1[illegible]—A scheme by the rural electrification administration [illegible] electric energy out of the atmosphere was described Su[illegible] Rep. Chas. A. Halleck (R), Indiana.

Halleck revealed the pr[illegible] a st[illegible] after reading testimony of the senate agriculture committee's investigation of the R E A.

He said the testimony showed R E A paid Thomas Henry Moray of Salt Lake City $25 a day to perfect a system of syphoning "radiant energy" from the air. The testimony said Feliz J. Frazer, an investigator for the works projects administration, spent two years "assisting, begging, cajoling and threatening Moray to develop his radiant energy."

Halleck said Franklin P. Wood, R E A engineer, was sent to see radiant energy in the making, and turned in a report which said, in part:

"Moray is unfathomable.

"In some inexplicable way he has, I believe, stumbled on to something that he himself cannot understand, but which, if it is all he thinks it is, will more profoundly affect the human race than any discovery in previous history."

Halleck said Moray was kept on the R E A pay roll after the engineer's report.

Moray worked from February, 1939, to February, 1941. ?? Not exactly true

These pictures were taken by Felix Frazer in April 1939 for the U.S Gov. during tests of the Moray Radiant Energy Radio. Meter shows reading of power coming from the Cosmos to operate the radio. Power from this radio was also sent thru 12 sheets of window glass to establish some of its characterictics. All these tests were made by Gov men long before the Gov signed the contract with Moray. Their signing proved their satisfaction in the results had at the tests. The Gov sought Moray's efforts. Moray did not seek the Gov contract.

See Moray statement of 1940-1941 on the communistic attitude of these Gov men

many other pictures were taken by Mr Frazer

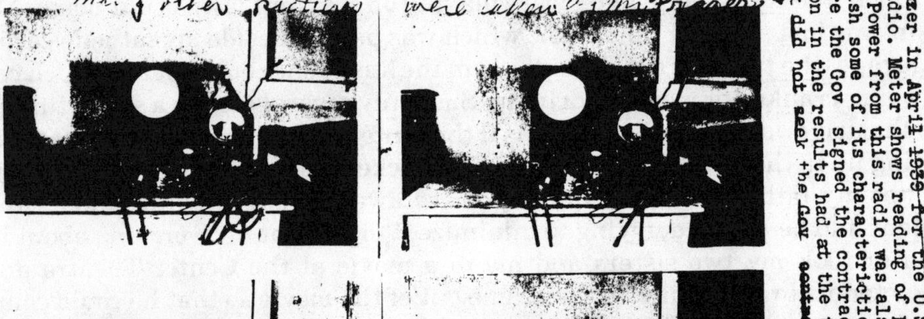

all sorts of tests were made And Moray Radiant Energy devices were on by: those making tests

glass plates — to radiant energy device — RE — to lights etc

current went thru the 12 pieces of glass as if they were not there.

Fig. 60

News clippings with comments by Dr. Moray

carrying a gun came running up. Dad kicked the second man, knocking free the gun he was carrying just as the first man's gun discharged. The bullet traveled vertically downward, grazing the side of Dad's leg as it went, and ricocheted off the concrete floor. At that point Dad's right arm began to function and he was able to get his own gun out. He gained control of the situation and showed the two men out the front door at gunpoint. He was immediately fired upon again by someone at a distance; he returned the fire, knocking the third gunman down. A fourth man rushed up to help the wounded gunman. Henry recognized this man as Felix Frazer. The second man said to the first assailant, "Well, you weren't as quick on the draw as you thought you were," and Henry Moray recognized the voice of an FBI man he had known at one time as a security guard. At that point, Henry realized he was all alone in a very dangerous and difficult situation. Here were two men, supposedly FBI agents, who in a shoot-out had gotten the worst of it. Henry was "severely wounded," he thought he was bleeding to death, and he knew he was going to faint at any moment. If he fainted while these men were still there, he knew that he would be completely at their mercy. So in panic he told them to get out, pretending that he had not recognized any of them, and the men promptly left. Henry locked the lab, went directly to the house and called his doctor and his good friend, S.E. Bringhurst. The doctor came and treated Henry's wounds, stopping the bleeding. The doctor knew Dad was doing government work, and Dad promised to report the incident to the FBI himself. Consequently, the doctor made no direct report. As far as I know, no police report was ever made.

It was Dad's belief that the continual harassment was to force him to turn over his notes to Felix and his friends. During this first attack, probably the four gentlemen only intended to abduct my father. Otherwise, why would they not have shot first instead of trying to hit him on the head?

The next morning, Sunday, my father asked me to go down and check the laboratory. He told me to go through my uncle's yard, taking my .22 rifle for my own personal protection. I was 13 years old. Upon reaching the west end of the garden, I was to watch the front door of the lab for half an hour to insure that no one was there; then I was to go to the door and check it to make sure that it had been locked, as he could not remember. I did exactly as I was told.

On Monday morning, we assisted my father to the laboratory at an early hour before anyone else arrived. Julius Noyes, his assistant at the time, was due to arrive at 8:00 a.m. Julius came in and went right to work in the back room after simply saying good morning, and Dad did not move from behind his desk. Later, Felix Fraser came in and rushed back to Julius Noyes. Shortly after, Fraser returned to the office and fussed around for a few minutes, looking at the floor. Then he came into my father's office and said, "Henry, why didn't you tell me you were shot?" Immediately Dad asked him how he knew that he had been shot. Fraser said, "Oh, Julius told me," but my father had deliberately prevented Julius from knowing of the shooting. Dad quickly changed the subject, but from that point on he refused any further cooperation with the Rural Electrification Administration.

Interestingly, Mr. Hans Amley, who was a faithful friend and bodyguard/

night watchman, had been relieved from his job just a few days before March 2, the date of the shooting. If Hans had been present on watch, Henry's attackers might have met an abrupt end, because Hans was noted for the fighting he and some of his friends had done in the Spanish Civil War.

Because of Dad's refusal to cooperate further with the REA after the shooting, the liberal backers from Southern California withdrew their financial support. Suddenly, a number of contractors were bringing actions against him to collect bills he thought had been paid months before.

Things began to move rapidly then. Dad refused to cooperate further with Felix. He discontinued any active direct correspondence with the government except through Bringhurst and his attorney, K.K. Steffenson. He insisted that the government-furnished equipment be taken out of the laboratory and that the government withdraw from the research.

In 1939-40, Henry was introduced by Dr. McManis to an English diplomat, M.V. Bluit, D.A.S. Mr. Bluit became very excited about Radiant Energy. In January of 1941, Mr. Bluit invited Henry Moray to come to England to work under the auspices of — we assumed from the correspondence — the English Admiralty. However, Dr. Moray felt that if he left the United States he would also leave behind what little security he had left. He knew it would be impossible for him to carry a gun in England. He also knew that he would have to transport his notes and materials. While I don't think he really believed that anyone would ever completely stop him from developing Radiant Energy, he had become more and more cautious and more afraid of the world as a result of the direct attacks upon him. Henry therefore declined the offer to go to England.

Henry's efforts thereafter were abruptly directed to surviving, physically and financially. He was given a fixed time limit to completely pay the debt he had incurred in building the laboratory. Funds became so tight that he could not even afford to have power in the building, and the laboratory appeared to be deserted. Henry became somewhat of a recluse and refused to plant lawn or shrubbery around the building. The yard was allowed to go to weeds. In spite of his precautions, on two other occasions after this he was attacked again.

In 1942, shortly after World War II began for the United States, Henry Moray attempted to rebuilt a Radiant Energy Device, using the remaining bit of what was known as the "Swedish Stone." This material, which was the heart of his original RE detector, he had never been able to duplicate, and the shortage of this material limited the amount of power he could draw. Consequently, in the larger unit, he developed a second detector that forced him into extensive research involving nuclear materials and radioactive reactions. He became deeply involved particularly in the study of synthetic radioactivity as described by Gustave LeBon in his book, **The Evolution of Matter.** He ran into a new problem at this point; he found that for some reason the book had been withdrawn from the public libraries. As a result of Henry's inquiries about the book, FBI agents came by and asked why he wanted the book. He told them, very plainly, "To read!" To Henry this investigation was harassment, pure and simple. He had no way of knowing that the Manhattan Project had started and the U.S. was deeply committed to a massive effort to develop the atomic bomb.

He had no way of knowing why a book more than fifty years old had been classified and was gone from every library in the United States. Consequently, his friend Wilson at the Wilson Book Store on 2nd South in Salt Lake City advertised for **The Evolution of Matter** and obtained three copies for him. One might ask why Henry needed three copies, but the answer is very simple. He had been accused before of falsifying and printing books for his own benefit by a "gentleman" at the University of Utah who had graciously given him credit far beyond his due in claiming that Henry had ghosted and reprinted one of Nicola Tesla's books for his own profit, calling it **Experiments with Alternating Currents of High Potential and High Frequency.** Henry always laughed at this preposterous charge. However, this is one of the reasons why he became a book collector and always collected more than one copy of any book.

At any rate, Henry Moray succeeded in rebuilding the small Radiant Energy device and began testing it for endurance in the summer of 1943. He put it in the back of his automobile, with a bank of lights as the resistive load, and drove the car with his family from Salt Lake City to Ashton, Idaho, into Wyoming, back to Salt Lake City, and to Denver, letting the device operate continually. In Greeley, Colorado, he took it out of the car, switched the device off and stayed overnight in a motel. The next afternoon he attempted to set up the device in a park in Denver. He had been promised backing by Dr. Spears, one of the originators of the Spears Clinic in Denver, and wished to demonstrate the device for him. I specifically remember how distressed he became at the time because after I had helped him set up the antenna and the ground, he was unable to tune the device properly and get it to operate. My father sent me away to get something to eat, and I grabbed a quick hot dog and returned. Dad was still quite upset. My mother spoke gently to him and said, "Henry, it's all right, come and get something to eat." I said to him, "Daddy, I will watch things." So he stepped over to the table to eat, and I sat and watched the "dead" bank of lights. After about ten minutes, the lights began to glow ever so dimly. As I watched, little by little the illumination built up and became distinct. I ran and got my father, and within a very few minutes the lights were burning very brilliantly, and it was obvious that the device was operating and in excellent condition.

But once again failure was to thwart his efforts, even with the new and successful rebuilding of the device.

He had been promised through a well-meaning second party that Dr. Spears would back the Moray research and insure that the RE device would be produced. However, when we left Spears' home that night we left in despair, for Dr. Spears' offer had been, to quote, "If you will leave your device here, I will contact you and let you know if I like it or not." Of course, this was intolerable.

This was the crowning and fatal blow to my father's research. We returned to Salt Lake City, somewhat fearful of what was about to happen. Henry had been given a time limit to either settle his debts or leave the laboratory and his home. Foreclosure procedures had already been completed. His reputation and his credit had been attacked.

But even this was not the worst. That Thanksgiving a man named Robert B. Perish, from Monrovia, California, came to Salt Lake. I personally heard him say to Henry, "Either come back to the government or we will liquidate you." I became so angry at this direct threat that I went to get my .22 caliber rifle to shoot the man myself, but my father stopped me in the back room — for which I am thankful to this day.

This brazen threat was the final blow to Dr. Moray's courage. Meanwhile, the RE device had burned out during a test in Salt Lake because of an overload in the circuit. The so-called "detector" was no longer operative, for out of fear of compromising its secret, he had dismantled it months before. He had been shot once and had exchanged fire with attackers on two other occasions, hitting several of his assailants and miraculously escaping with his life. Now it appeared that the U.S. Government was saying that he was to be liquidated.

However, Dad countered by telling Mr. Perish that the RE notes had been divided into separate packages and sent to secret friends in England and Canada. He stated that these notes were hidden in safe places with the instructions that if anything ever happened to Henry Moray, the notes were to be published immediately and in full. Perish then replied that Henry would never be bothered again.

As a result of Perish's visit and delivery of a threat of liquidation by the Government, Dr. Moray began to refuse to see anyone connected with the Government. He buried himself completely in the therapeutic research. From time to time there would again be evidences in his laboratory work of his RE effects. However, he felt that because of the high cost of materials and of gaining legitimate assistance, it would be almost impossible to rebuild the device.

Because Henry himself did not have powerful financial backing, most individuals refused to do any investigation of the RE device and its operation from a simple description unless they were hired and paid.

The years slipped by and from time to time Henry and I would discuss his notes and go over the drawings. He spent most of his time working on what he called the "counter-balance" to eliminate the need for an antenna. As time went on, I became afraid the antenna would fall. The neighborhood was becoming quite built up. A seventy-foot pole falling on a neighbor's house would have done considerable damage, so I had the antenna dismantled in 1956.

In 1949-50, my father, my brother, and I had a conference and discussed what could be done to raise sufficient capital for RE development. My brother thought Canada would grow, and proposed going there and investing in land to eventually raise money as the land increased in value. I volunteered to stay in Salt Lake City and work with my father. Consequently, while we continued to try to find other research projects to support our research in Radiant Energy, we divided our forces. My brother went to Canada, and my father and I stayed in Salt Lake City.

Needless to say, through the years my father and I were not very successful in raising capital because most of the projects we obtained either barely paid their own expense plus a living wage, or they required additional capital to support them.

**Note made by Dr. Moray of this visit reads: "Robert B. Perish, 718 Mountain View Avenue, 5944 Monrovia."*

I became interested in high energy radiation effects and spent much time investigating the effect of high energy electron and photon beams on materials. I received recognition for my work when I delivered a paper in 1964 in Colorado before the Colorado Mining Association's convention. I have been listed in **American Men of Science** since that time.

My brother's work has been successful financially, and he has supported much of the research the last few years. My father was an innovator rather than a theoretician in his experimentation, and consequently, his notes are more general than specific; but my brother and I believe that by combining our personal knowledge with his notes and other documentation we would be able to rediscover — and with financial backing, rebuild — the RE device. Though it may cost a fortune, we do not intend to allow his work to remain buried.

Radiant Energy can no longer be stopped by simply ignoring it or attacking one or two individuals, for we have been able to build an organization where several persons are qualified to carry on the work. The only remaining difficulty is that this type of research and development takes several millions of dollars and long periods of time. Thus financial assistance is still needed, and assistance is needed from individuals willing to work without immediate remuneration towards the salvation of mankind in this increasingly energy-short world.

In the past, Dr. Moray never intended to make large sums of money for himself. He only wished to insure that the device be made available to mankind instead of being hidden or used by a few powerful men for a monopoly. I think this can best be proven by examining the record of his work with the Judd interests of San Francisco. I have two letters which are directly applicable, one from his attorney, K.K. Steffenson, and another one from Mr. Edward Sasoon. Both persons were present when Judd of San Francisco offered Dr. Moray one hundred million dollars for a 51 percent interest in Radiant Energy. Dr. Moray refused because selling a 51 percent interest would mean passing control of Radiant Energy into the hands of men who would then never offer it openly to the world.

On October 16, 1956, Dr. Robert Craig, a former REA deputy administrator wrote: "The development of Radiant Energy has been a slow, laborious task, one of the greatest difficulties being that Moray has stood all alone. People who should have been interested were more concerned with immediate returns than with the important long-range returns. Naturally, during such a long period, many questions arise as to protection of rights, interests, etc."

Again on November 14, 1959, Dr. Craig wrote, "Permit me to say that I have known Dr. Moray now for about 20 years and have observed him single handedly trying to get acceptance of some of his ideas and discoveries, particularly in the area of 'Radiant Energy.' Many of the now accepted areas in atomic and nuclear physics were outlined by Dr. Moray as early as the middle 30's.

"At one time or another I have endeavored to help him by bringing his work to the attention of people in the government without any great success. Of course, no longer do they dispute the basis of his work. However, to bring pure research to successful commercial application involves time and money, particularly the latter. While we can spend millions for missiles that are abortive and no one seems to mind, a few hundred thousand dollars available to this man would obviate, I believe, much of the need for missiles. Of course, this is my own idea in the matter, but I do believe a helping hand now to Dr. Moray could do more towards solving world tension than the mere building of bigger and bigger armaments to protect us from the 'have-nots.'

"I can assure you from my personal knowledge of Dr. Moray that here is a unique mind and an ability which would be worthwhile developing."

Henry Moray became more and more amazed as he approached death, for he had never believed he could really be stopped. The words of Perish had essentially come true. He had been liquidated. Many people in the world had attacked his credibility, many of his records had disappeared, and the true background of things that had happened to him was generally unknown. In fact, we have recently been informed that none of the original patent applications that Henry made are any longer available at the U.S. Patent Office. Although their file jackets are there, the contents and applications themselves are gone. Methodically, more than a dozen applications have "disappeared." The question of exactly who has removed them will probably forever remain unanswered. But I can assure the reader that Watergate was not the first great cover-up and act of duplicity — and it will not be the last.

Others may discover direct energy conversion systems using synthetic radioactive materials. Bell Laboratories has made millions of dollars from semiconductors though no credit has been given to Henry Moray. Therapeutic or chemical reactions that Dr. Moray worked with may be extensively used in the next few decades.

I knew and had the privilege to work with one of the most brilliant scientific minds that ever lived who received little credit. Using, as a friend of his calls it, "second and third order of magnitude thinking," Henry Moray produced free Radiant Energy that could lift mankind from his servitude and energy limitations and place the very stars in his grasp. Even though I am his son and may be accused of being prejudiced, no one who looks at the record can say that my claims or my feelings are entirely unjustified.

INTRODUCTION TO SECOND HALF OF THE BOOK

The following chapters were originally written by Dr. Moray; however, we have edited the original **Sea of Energy** to make the work read more smoothly. The information here was originally published by the Moray Products Company in a book called **Beyond the Light Rays,** which is out of print and no longer available. The last known copy was stolen from Dr. Moray's laboratory in the late 1940's by a dentist "frienemie."

CHAPTER 7

THE SEA OF ENERGY IN WHICH THE EARTH FLOATS*

By T. Henry Moray, D. Sc.

A NEW ERA IN ENERGY: Power from the Cosmos and the Earth.

Dr. Nikola Tesla said over sixty years ago:

> Ere many generations pass, our machinery will be driven by a power obtainable at any point of the universe . . . throughout space there is energy. Is this energy static or kinetic? If static our hopes are in vain; if kinetic — and this we know it is, for certain — then it is a mere question of time when men will succeed in attaching their machinery to the very wheelwork of nature.**

Nikola Tesla was not referring to atomic or nuclear energy, but to the energy which is continually bombarding the earth from outer space.

Enough energy is coming to the earth to light over 1.5 million (1,693,600) 100-watt lamps for every human being on the earth today.*** No fuel of any kind need be taken as a dead load as this energy can be "picked-up" directly by ocean liners, railroads, airplanes, automobiles or any form of transportation. Heat, light and power can be made available for use in all kinds of buildings and for all kinds of machinery. An example would be to pump water onto the desert lands, the power source being only a fraction of the weight of any steam plant or any kind of engine in use today and all this at a fraction of the current cost.

A wild dream? No! It's a proven practical reality, as hundreds of people know who have witnessed the Moray Radiant Energy invention — powered from the cosmos. This is cosmic energy.****

The total energy involved in "cosmic" radiations is very large. The mechanism of its generation involves a basic relationship with the total structure and action of the universe. Today, it is believed that cosmic radiation consists primarily of protons and some heavier nuclei. At times this cosmic energy packs a wallop of around 100 quadrillon electron volts. Coming continuously with slight variations in time, the radiations have a uniformly directional isotropy.

**This was Dr. Moray's original chapter 1*

****Nature of World and of Man,** *Chicago Institute of Technology*

*****Experiments with Alternating Currents of High Potential and High Frequency,** *Nikola Tesla, 1904.*

*****Dr. Moray was not refering to Cosmic Rays. One must read further and contemplate his hypothesis regarding energy from with the atom.*

The earth is, therefore, surrounded in an atmosphere of radiation with cosmic rays coming continually to the earth from all directions, although there may be a slight deflection of the weaker rays by the earth's magnetic field. There is every indication that our sun is not the source of any appreciable amount of this radiation. The origin, therefore, is from the universe as a whole. The total energy of cosmic radiation is more than the entire luminous output of all the stars and nebulae of the universe combined. Unlimited power is being delivered to everyone's doorstep.

The Moray Radiant Energy discovery, using radiations from the cosmos as its power source, gives the greatest amount of energy per pound of equipment of any system known to man. Electrical power through an electric motor or an electric jet far exceeds any form of energy in any engine in the delivery of power. There is no dead center of lost motion in an electric motor nor loss of push in an electric jet. Also, the starting torque is much higher in the electrically powered engine than in the combustion engine.

Harnessing cosmic energy is the most practical method yet discovered by man. Furthermore, it is possible to utilize this vast source of energy from the universe without a prime mover at any point on the earth — on the ground, in the air, on the water, under the water, or even underground. If one considers that an electrical generator is not in the true sense a generator — as electricity is not made by the generator — but is merely an electrical pump, the Moray Radiant Energy device may then be referred to as a cosmic ray pump: that is, a high speed electron oscillator serving as a detector of cosmic radiations which causes a pumping action or surging within its circuitry.

To account for the propagation of heat and light — two of the forms of Radiant Energy — man has postulated the existence of a medium filling all space. But, the transference of the energy of radiant heat and light is not the only evidence in favor of the existence of such a medium. Electric, magnetic, and electromagnetic phenomena and gravitation itself point in the same direction.

Attractions and repulsions take place between electrified bodies, magnets, and circuits conveying electric currents. Large masses may be set in motion in this manner, acquiring kinetic energy. If an electric current is started in any circuit, corresponding **induced** currents spring up in all very closely neighboring conductors. Yet, there is no visible connection between the circuit and the conductors. To originate a current in any conductor requires the expenditure of energy. How, then, is the energy propagated from the circuit to the conductors? If we believe in the continuity of the propagation of energy — that is, if we believe that when it disappears at one place and reappears at another it must have passed through the intervening space and, therefore, have existed there somehow in the meantime — we are forced to postulate a vehicle for its conveyance from place to place.

When a particle is electrified, what one must first observe is that a certain amount of energy has been spent; work has been done. The result is an electrified state of the particle. The process of electrifying a conductor is, therefore, the storing of energy is some way in or around the conductor in some medium. The work is spent in altering the state of the medium, and when the particle is discharged, the medium returns to its original state, and the store of energy is disengaged. Similarly, a supply of energy is required to maintain an electric current, and the phenomena arising from the current are manifestations of the presence of this energy in the medium around the circuit. It used to be that an electrified particle or body was supposed to have something called "electricity" residing upon it which caused electrical phenomena. An electric current was regarded as a flow of electricity traveling along a wire (for example), and the energy which appeared at any part of a circuit (if considered at all) was supposed to have been conveyed along the wire by the current. But, the existence of induction and electromagnetic interactions between bodies situated at a distance from each other leads one to look upon the medium around the conductors as playing a very important part in the development of these electrical phenomema. In fact, it is the storehouse of the energy.

It is upon this basis Maxwell founded his theory of electricity and magnetism, and determined the distribution of the energy in the various parts of an electric field in terms of electric and magnetic forces.* The medium around an electrified body is charged with energy and not of an imaginary electric fluid distributed over the electrified body or conductor. When we speak of the charge of an electrified conductor we are referring to the charge of energy in the medium around it, and when we talk of the electric flow or current in the circuit we are referring to the only flow we know of, namely, the flow of energy through the electric field within the wire.

The work in producing the electrification of a conductor is spent on the medium and stored there, probably as energy of motion. To denote this we shall say that the medium around the conductor is polarized, this word being employed to denote that its state or some of its properties have been altered in some manner and to a certain extent depending on the intensity of the charge. If the charge is negative the polarization is in the opposite sense, the two being related, perhaps, like right-handed and left-handed twists or rotations.

Now consider the case of a body charged alternately, positively and negatively in rapid succession. The positive charge means a positive polarization of the medium, which begins at the conductor and travels out through space. When the body is discharged the medium is once more set free and resumes its former condition. The negative charge now induces a modification of the medium or polarization in the opposite sense. The result of alternate charges of opposite sign is that the medium at any point becomes polarized alternately in opposite directions, while waves of opposite polarizations are propagated through space, each carrying energy derived from the source or agent supplying the electrification. Here, then, we have a periodic disturbance of some kind occuring at each point, accompanied by waves of energy traveling outwards from the conductor.

The phenomenon of interference leads to the conclusion that light is the result of periodic disturbances or vibrations of the medium, but as to the nature of

*__Atomic Physics,__ *Physics Staff University of Pittsburgh, 2nd Edition Ch. VIII, John Wiley & Sons*

these vibrations, as to the exact nature of the periodic changes or what it is that changes them, we possess no knowledge. We know that alternating electric charges are accompanied by corresponding changes of state or vibrations of the medium, and if the charge if varied periodically and with sufficient rapidity, we have a vibration at each point analogous to, perhaps identical with, that which occurs in the propagation of light — a combination of wave and particle properties. This then is the electromagetic theory of the luminous vibration.

In the older elastic-solid theory, the light vibrations were supposed to be actual oscillations of the elements or molecules of the medium about their positions of rest, such as takes place when waves of transverse disturbance are propagated through an elastic solid. Such limitation is unwarranted to some extent, but one cannot afford to entirely disregard the particle theory of light either. A combination of the theories has merit. We know that the change, disturbance, vibration, polarization, or whatever we wish to term it, is periodic and transverse to the direction of propagation. The electromagnetic theory teaches us nothing further as to its nature, but rather asserts that whatever the change may be, it is the same in kind as that which occurs in the medium when the charge of an electrified body is altered or reversed. It reduces light and heat waves to the same category as waves of electric polarization. The only quality of the latter required to constitute the former is sufficient rapidity of alteration. These speculations were given the strongest confirmation by experiments of Professor Hertz* many years ago.

When a resilient substance is subjected to strain and then set free, one of two things may happen. The substance may slowly recover from the strain and gradually attain its natural state, or the elastic recoil may carry it past its position of equilibrium and cause it to execute a series of oscillations. Something of the same sort may also occur when an electrified capacitor is discharged. In ordinary language, there may be a continuous flow of electricity in one direction until the discharge is completed, or an oscillating discharge may occur. That is, the first flow may be succeeded by a back-rush, as if the first discharge had over-run itself and something like recoil had set in. The capacitor thus becomes more or less charged again in the opposite sense, and a second discharge occurs, accompanied by a second backrush, the oscillation going on until all the energy is either completely radiated or used up in heating the conductors or performing other work.

When capacitors are filled with energy captured by the Moray Radiant Energy device and then discharged through a circuit of proper impedence, reactance and inductance, thereby syncronizing the oscillations of the device with those of the universe,** electrical inertia is set-up. In the reversal of the current, the capacitors are charged, discharged and recharged slowly until the energy stored in them is radiated in kinetic energy through the device, and this

***Atomic Physics,** *Physics Staff University of Pittsburgh, 2nd Edition, Ch. VIII, John Wiley & Sons*

***Very often Dr. Moray interchanged the word universe with matter.*

energy can be kept alive indefinitely by establishing resonance with the oscillations of the universe.

Considering oscillations from a mechanical, electrical and mathematical point of view, we find that electrical resistance is the same as mechanical friction and current is comparable to mechanical velocity. Inertia and inductance then may be considered analagous terms. In mechanics the greater the inertia of a body, the longer it will stay in motion. In the Radiant Energy device's resistance-inductance-capacity (REC) circuit, the greater the electrical inductance, the longer the current continues to flow once it is established by synchronization with cosmic surges.

Expressed mathematically, the equations are the same for electrical or mechanical phenomena. That is, $R < \sqrt{4L/C}$, where R is the resistance in ohms, L is the inductance in henries, and C is the capacity in farads. When this is true, an oscillatory discharge will occur and a very powerful inductance inertia will assert itself. For low values of R, the frequency of the oscillations can be shown by $f = \frac{1}{2}\pi\sqrt{CL}$. The rapidity of the oscillations' frequencies are governed by the capacitance and inductance.

In the vibrational forces of the universe, we find the key to the source of all energy. How we can utilize this energy for modern industry without being limited to mechanical prime movers is the question. And, the answer may lie in an energy generator, balanced so as to oscillate in synchronization with the oscillations of the universe.

Dr. Ross Gunn, a civilian scientist for the U. S. Navy, stated years ago that the earth is a huge generator, generating over 200 million amperes of electric current continuously. For example, the aurora borealis is considered to be a very definite electrical phenomenon produced by the passage of electric charges through the rarified gases of the higher atmosphere. The earth has since been shown, by Dr. Gunn and others, to have a negative charge amounting to 400,000 coulombs. Yet, six feet above the ground the air is charged with more than 200 volts — positive with respect to the ground.

It is known that air conducts electricity away from charged objects. This being true, how does the earth, a charged object and exposed as it is to the surrounding atmosphere, maintain its charge? If the air conducts electricity, the earth's charge must be constantly passing into the atmosphere. And it has been calculated that the earth has a continuous discharge into the atmosphere of 1800 amperes. At this rate, the earth would lose 90% of its charge into the air in one hour, yet the earth's charge does not diminish. Where does the earth's source of energy come from?

The conversion of matter to energy in the stars is accepted, and, reasoning from what occurs in radioactive disintegration during which energy waves are radiated, one may conclude that energy waves of very high frequency are sent out from the stars (one of which is our sun). Now, of course, the conversion of energy into matter must equally be accepted.

It has been found that ionization, which could be the medium for the flow of energy, increases with altitudes instead of decreasing. Since the source of energy is the universe, the generation of energy by rotary action and by all prime movers is an effect and not a cause. Oscillatory energy action, be it in a Leyden jar or another man-made capacitor, or in what we may call natural capacitors, behaves the same. The oscillations will continue until they have reached their cycle of height and then there will be a back-rush returning to where the oscillations originated. Every oscillation, whether large or small, is completed during the same interval of time. These oscillations all prove the same great fact, that they are governed by the same cycle of time, completed during the same interval of time. Waves of energy have a regular beat note, coming and going as the waves of the sea, but in a very definite mathematical order — coming to the earth from every direction with a definite rhythm.

Energy has a definite elastic or resilient rigidity and density, which is subject to displacement and strain. When strain is removed, the medium will spring back to its old position and beyond, surging back and forth, and will continue to oscillate until the original pressure is used up. If the internal impedance is too great, there will be no oscillations, but it will merely slide back in a dead beat to its unrestrained state.

By cutting down resistance to a minimum and synchronizing the resilient ionic actions of the Moray device with the wave actions of the universe, periods of oscillation can be made to come quicker and quicker until inertia asserts itself, thus lengthening out the time of final recovery. This is done by carrying the recoil beyond the natural oscillations and prolonging the vibrations by capturing them in oscillatory action. When the recovery becomes distinctly oscillatory, a harmonic pattern is initiated and the oscillations continue, resonance thereby being established with the universe.

In the universe we see the same laws being obeyed as in our laboratories. As one traces down to the almost infinitesimal constituents of the atom, one finds that matter does not exist at all as the realistic substance which we have supposed it to be. There at the very foundation, it consists of nothing more than energy charges emitted at various wave lengths or frequencies. It is becoming more and more certain that the apparent complexity of nature is due to our lack of knowledge. And, as the picture unfolds, it promises a marvelous simplicity.

One of the most marvelous relationships that has ever been revealed in the entire science of physics is that between light and electricity and the existence of electronics in atoms of matter. Knowing what we do at the present time with regard to the structure of atoms, this relationship is not quite so surprising. However, considering the total absence of this knowledge about a half century ago, the discovery that light, and radiation in general, are vibratory phenomena was revolutionary.

Speaking of radiation, Radiant here means proceeding from a center in straight lines in every direction. Energy is internal and inherent. "Energy is defined as a condition of matter in virtue of which any definite portion may affect changes in any other definite portion." This was written in 1892, and discoveries since confirm it. Energy then is a state of matter, or rather, the result of a particular state or condition in which matter may be when any observed phase of energy appears.

In addition to possessing kinetic energy, the atom is capable of absorbing energy internally. This internal energy is associated with the configuration of the particles of which the atom is composed. Under ordinary conditions an atom is in what is known as a state of equilibrium in which there is neither a giving off nor absorbing of energy. But, the internal energy of the atom can be altered. When the internal energy of the atom exceeds that of its normal state it is said to be excited. Excitations may be caused in several ways, e.g., the collision of an atom with rapidly moving positive or negative particles or the breaking of lines of force in an electromagnetic generator. Kinetic energy is released when excitation causes a particle to give up some or all of its kinetic energy to the atom during collisions. This is taking place in the universe all the time.

The electric motor and generator would never have been discovered if a dielectric (insulation) had not been discovered. If one discovers a dielectric valve for the energy of the universe and a means of making this device oscillate with the oscillating energy of the universe, one has the answer to harnessing the energy of the universe! A limiting case of excitation is ionization, wherein energy is absorbed by the atom sufficiently to allow a loosely bound electron to leave the atom, against the electrostatic forces which tend to hold it within the atom. An atom which has given up one or more electrons is said to be ionized. It is possible that ionization, i.e., excitation, may take place in successive steps through absorption of quanta energy. The return of an ionized atom to a state of lower energy is associated with electromagnetic radiation. Also, from the process of ionization, electrical energy may become associated with the vibrational forces of the universe coming into the earth as cosmic radiations. The higher the frequency, the greater the ionization or excitation, a form of energy which is kinetic in nature. There are tremendous energies coming to the earth from outer space. These energies are only different manifestations of the energies we see in operation all around us. In most cases we are not even aware of their existence. "They penetrate everything including our own bodies. Everyone of us is alive by virtue of these energies. Every part and particle of the universe is alive with them. The generators that now furnish our electric power do not create or originate any power or electricity; they merely direct, **pump,** the existing energy or electricity."*

As in musical notes of high and low "C", the vibrational rates (frequencies) are different, but all "C" notes are essentially the same (harmonically related). This is the foundation upon which much of my investigation of vibratory phenomena is based.

It has been agreed that all forms of matter are vibrating at a particular rate of frequency. And, so it is with the various forms of energy — heat, light, magnetism and electricity. These are but forms of vibratory motion connected with and being generated from the same source, the universe. Matter vibrates at a particular rate, according to its character, and may be transmitted into other substances by lowering or raising its rate of frequency. If the frequency is raised high enough, the molecules will separate and the atoms become free. Raising the frequency still higher, the atoms resolve themselves into their original

***Advance of Science,** *Watson Davis, 1934*

components. Matter then becomes a form of energy. Frequencies may be developed which will balance the force of gravity to a point of neutralization. One can then go beyond the force of gravitation. Understanding the principles of vibration is truly understanding energy.

In gamma rays, we find potentials which are equivalent to as much as 2,000,000 volts, yet their wave lengths are not the shortest known. In octaves still higher lie rays which are known as cosmic rays. Who can draw a definite line and say how much higher other octaves exist than those known as the cosmic rays? Our starting point from the discovery of these different waves was electrical conductivity of the air, and it has been found that this conductivity is just as strong by night as by day. Radiations emitted by the sun can scarcely be the sole cause of this energy. All space is saturated with vibrations, energies, which are no doubt electrical in character. The relation of matter to energy and energy to matter then becomes the potential of the universe — one continuous series of oscillations.

Atoms maintain an equilibrium by oscillations, rotations, attractions and repulsions, but this does not interfere with a transformation of equilibrium, which, when the transformations of equilibrium are rapid enough, become energy, i.e., matter is turning into energy and energy into matter.

There can be no generation of electrical current and no kinetic energy if there is no disturbance of equilibrium, i.e., change of potential or change of energy levels. When one thinks of the oxygen and nitrogen molecules of the air all about us moving with the speed of bullets and striking us and everything else at this speed, one can form some idea of the agitation taking place here and in the universe.

The oscillations from outer space are emitting electromagnetic waves of many wave-lengths and frequencies. The Moray device is so constructed that the frequency is very much lower on the secondary side than on the primary side, and almost complete resonance is established. I am convinced the energies from the universe are active radiations produced by the evolution of matter into energy and energy into matter.*

Dr. Anderson's cloud chamber at California Institute of Technology in which the positron was discovered has furnished much information about cosmic ray energies. He found that some positrons are born of cosmic rays smashing into matter. The cosmic ray energies deduced from the tracks left in the Anderson cloud chamber range from 100 volts to 3 billion volts. The Lemaitre-Vallarts theory together with Dr. Johnson's asymmetry measurements, give definite values for the energy of half of the cosmic radiation, and shows it continuously distributed between 5 billion and 50 billion volts.

The figure of 100 billion volts is a result of Dr. W. Kolhorster's measurement of penetrating radiation in the depths of the Strassfurt salt mines. He found that the minimum energy of these rays had a penetration which was greater than

(At this point Dr. Moray observed the discoveries of the early agents to be paralleling some of his own ideas. Then he went on to quote several past releases of the day (1960).

ever before demonstrated. Dr. Axel Corlin of Sweden's Lund Observatory found radiation that still had energy after passing through somewhat greater depths and, therefore, the voltage figures can be made even higher. Energies of 100 billion volts or more are indicated by the great bursts set off by cosmic ray collisions, called the stosse, which have been observed particularly in Germany. The Moray RE devices have worked equally well in deep mines, under water or high in the mountains and in an airplane.

It is about 100 years since science began to consider light, heat, magnetism, galvanism and electricity as natural forces. In the early part of the 19th Century school books termed these things "imponderable substances." The corpuscle theory of light was taught, the sun was supposed to provide an endless supply of these corpuscles. After the corpuscle theory faded, scientists turned to the wave theory, but even that was based on a crude concept of movement of the ultimate principles or atoms, of matter. The electron theory has superceded the earlier ones now, and while the electron theory explains the observed and theoretical "facts" better than the previous concepts did, could it be that, as the greater light of knowledge leads us on, the electron theory in turn will fall short of providing "absolute" knowledge? The Einstein Theory may stand in need of revision or amendment; or, in time, it may join the theories of corpuscles and waves on the back shelf.

CHAPTER 8

REACTIONS BY MEANS OF ELECTRON EXCITATION

As I stated in 1914 in connection with the action of Moray Radiant Energy tube values, if two molecules are beyond each other's molecular range and if the neighboring surfaces can by any means, as by the supply of electricity from without, be oppositely electrified, the forces of cohesion are intensified momentarily by a bonding effect or a polarization, and this force encompasses large molecular distances. The opposite charges cannot be maintained, but they can be held temporarily by imparting a sudden electric impulse. This is an effective method of promoting chemical cohesion. The electrons in two polarized molecules need not be disturbed to any great extent in order to reach this state of excitation. Ordinary molecular forces can thereby be temporarily pushed into a state of chemical imbalance.

Electrical forces between charges are of the inverse-square type, and the interactions of two charged particles free in space can be easily computed. Coulomb's Law states that the force between two small charged bodies varies directly as the product of their charges and inversely as the square of the distance between their centers. This force is also affected by the type of medium in which the charged bodies are placed.

Thus $F = \frac{q_1 q_2}{4\pi p r^2}$ where p is the permeability of the medium.

As an illustration of the force measured by Coulomb's Law, consider that we have two small bodies, each charged with one coulomb of like charge of electricity, q_1 and q_2, and that these bodies are placed in a vacuum at a distance of one meter. The force of repulsion will then be 9.0×10^9 newtons or two billion pounds of repulsive force.

Molecules can be made to obey Coulomb's Law and may be subjected to inter-molecular forces, and these electrical forces between reacting molecules are very large.

Gaseous ions exert small forces on each other at standard pressure and temperature at a distance between their centers of 10^{-2}cm and even as close as 10^{-5}cm. These ions move about randomly as a result of impact with other molecules (Brownian motion), and the forces of impact are much larger than the electrostatic attraction. Only when they are within a certain distance of each other is the coulomb potential energy equal to or exceeding the average of translation.

Thus, $\frac{e^2}{4\pi p r_o{}^2}$ is equal to or greater than $\frac{3Kt}{2}$, where r is the molecular distance.

Inside this molecular distance, the ions are drawn together; outside this sphere of radius r there is an electric field given by the equation, $X = \frac{e}{r_o^2}$

Thus, at 10^{-8}cm., X is $\sim 10^9 \frac{\text{volts}}{\text{cm}}$, and even at 10^{-6}cm., it is $\sim 10^{13} \frac{\text{volts}}{\text{cm}}$. It is

apparent that the coulomb force between oppositely charged ions that are within such molecular distances is large. The effect is analogous to the entrapment of a comet when it passes near a planet, the attractive force rendering it a permanent member of the solar system. The stoppage of an ion during an encounter with an oppositely charged ion occurs well within the limits of atomic magnitude (10^{-8}cm), so that the acceleration will be of the order of $\frac{U}{2t}$. The force needed to stop even a single electron will be 1/10 dyne. The power to stop and neutralize such electrons flying at 1/30 the speed of light inside a molecular thickness can be estimated as follows:

$$\frac{\text{Energy } u}{\text{time}} = \frac{\frac{1}{2}mu^2 u}{2t} = 10^{-27} \bullet (10^9)^3 \bullet 10^8 = 10^8 \text{cm./sec.}^*$$

Editor's note:

One may ponder what Henry Moray meant by this formula when he discusses the energy per second times the velocity. In the fourth edition of his* **Sea of Energy *book, on page 29 of chapter 2, an error was made in proof-reading, wherein (1) a "U" was left off the original energy over time formula and (2) the minus sign in front of the exponential 27 was inadvertently left off, which would give the answer of 10^8 ergs per second times the velocity. My experience with Henry Moray is that he very often used the abbreviated mathematical procedure, not showing all his steps. I would suspect in proofreading the 4th edition he would have double checked only the final answer for 10^8 ergs per second. In analyzing the mathematics he used, it is apparent that he developed his hypothesis by first exploring the Brownian movement, then the energy released, due to the capture of an electron, going on into Bohr's theory along with others, and then culminating with resonant reactions (in this book under chapter 8). It is my conclusion that Dr. Moray is equating the energy released by the atom or matter, as you wish, recognizing the fact that it is not an obscure mathematical condition wherein the atom is being held at a defined point, but in reality that all matter is moving and therefore the energy released is related to the velocity of that moving matter. If one compares this analogy to the stars or even our own universe, then in comparing the earth to an electron, we must realize that the energy released when our earth was captured by our sun must have been tremendous. Moreover, our solar system or sun is moving and the total energy is related to the total velocity of the solar system. If you will refer to the last page, chapter 7 of this edition, you will see that Henry Moray is analyzing mass relationship, and that this energy released is not necessarily the energy of deterioration, but the energy of generation. Henry Moray continues to analyze this into the oscillations of the universe, and oscillations of the electrical system. Take into consideration the modern hypothesis of the Neutrino sea. (See "The Neutrino Sea - Hypothesis or Reality?"* **Industrial Research,** *Dec. 1977, p. 51)*

Look at the research in chemistry of the discovery of salvated electrons. Both theories show that electrons and neutrinos are continually building up around us. One can then speculate, with the continual movement of everything around us and the building up of energy levels, that J.A. Wheeler and Henry Moray are talking about the same thing. J.A. Wheeler explains: that a close look at Einstein's explanations with regard to the geometry of electromagnetism and the correct connection between electricity and topology, gave rise to the fact that in 1955 and 1957 it was discovered that mass and electricity can be fashioned out of curved empty space. "A closer look at it by Misner in 1956, gave rise to the proposition of electromagnetism as a property of curved, empty space." Wheeler went on to say.*

"Rainich already long before in an almost forgotten paper had shown under what conditions a curvature of space - time can be regarded as due to an electromagnetic field. Einstein's 1916 theory can be regarded as an already unified-and entirely geometrodynamical - theory of electromagnetism and gravitation.

"Already in 1955 it had been found that out of electromagnetic radiation or gravitational radiation, or any mixture of the two, one can in principle construct a gravitational-electromagnetic entity, or geon. A geon though built of curved empty space and completely free of any so-called real mass, is nevertheless endowed — by reason of its radiation content, with mass of its own.

"A completely geometrical model was found at the same time for electricity, as lines of force trapped in the topology of a multiply connected manifold.

"Due in good measure to the physical insight and mathematical knowledge of Charles Misner, it was possible two years later, in 1957, to trace out the beautiful relations between this topological description of electricity and a central topic in modern mathematics, the theory of homology groups.

"There is not the least direct connection between the geons of classical geometrodynamics and the elementary particles of the real world of quantum physics. Also this theory in its classical version does not constrain the charges with which it deals either to be quantized or to be all of the same complexion - all electric, none magnetic as observed in nature. These circumstances hardly deprived geometrodynamics of physical interest.

"General relativity was already before 1955 a kind of master theory - the only description of physics to connect gravitation, geometry and the equivalence principle, and the only theory out of which the equation of motion of a mass can be derived as a consequence rather than being fed in as a postulate. Geometrodynamics retains these features and provides in addition a geon model for mass and a topological description for charge. Remote as are both pictures from anything of direct relevance to experiment, they indicate that curved empty space is a building material of previously unsuspected richness.

***Geometric Dynamics,** *by J.A. Wheeler, printed by Academic Press of New York and London.*

***"Neutrino Structure of the Ether", by M. Ruderfer, Dimensions, Inc., Hempstead, N.Y. "Lettere Al Nuovo Cimento", Vol. 13, N. 1,3 March, 1975.*

"It is natural to turn from a physical theory of such scope to still deeper questions. Are all fields and particles constructed out of curved empty space? Does there exist some high and subtle and uniquely right theory of point sets out of which some day can be derived the quantum principle; the approximate validity of **four**-*dimensional Riemannian geometry; the Rainich-Misner equations; neutrinos; and elementary particles? Advances since mid-century give new interest to these questions. Experiments at high energies have brought to light remarkable symmetries between the elementary particles. The discovery of parity non-conservation has revealed a connection between electricity and geometry—between the charge of a nucleon and the right or left handedness of the electron which it emits in a beta process.*

"Geometrodynamics is very far from being able to contribute to a discussion of these questions. It has nothing directly to say about meson theory or the origin of nuclear forces. Quantum geometrodynamics speaks if at all to a point closer to the foundations of physics. It advises that any multiple connectedness is a property, not of elementary particles, but of all space. Moreover, everywhere in space, quantum fluctuations in geometry and in electromagnetic fields are calculated to have such high energy density that the extra energy density associated with the presence of an elementary particle is negligible by comparison."

I am convinced that Henry Moray has verged on an explanation of the generation of energy, and through his "tubes" (semi-conductive devices mounted in glass under ideal atmospheres), he demonstrated that this energy does exist.

One must realize how frustrated Henry was because of the limitations of his background. Consider for an instant the word limitation of Democritus in the light of present knowledge when he wrote his explanation of matter 400 years B.C.*

"Atoms are infinite in number and infinitely varied in form. They strike together and the lateral motions and whirlings are the beginnings of worlds.

"The varieties of all things depend upon the varieties of their atoms in number, size and aggregation.

"The soul consists of fine, smooth, round atoms like those of fire. These are the most mobile of all. They interpenetrate the body, and in their motions the phenomenon of life arises."

The reaction of excited atoms or molecules is of importance only under conditions of high electric intensity in view of the short time intervals involved (10^{-8}sec.). Some atoms can, however, have electrons in a metastic state of excitation lasting some 10^{-4}sec. However, at lower electron densities second impacts can change the phenomena, and such atoms in impact with neutral atoms or molecules of an appropriate sort can lose energy by inelastic impacts causing excitation, ionization or dissociation of the molecules.

**"An Outline of Atomic Physics," by members of the Physic's Staff, University of Pittsburgh; 2nd edition, 1942.*

In any collision between a charged particle and a neutral molecule, ionization takes place because of the electric force exerted on the planetary electrons in the molecules.

The Bohr Theory of spectral lines indicates that an electron should be able to lose energy to an electron in an atom or molecule as soon as it possesses an energy equal to $hv = \frac{1}{2} mv^2$, where v is the frequency of the energy radiated, h is the Planck constant. Therefore, we have $h = 6.62 \times 10^{-27}$ ergs/sec., where v = frequency, m = mass. When the disturbed electron returns to its orbit or state, it was suspected that the first inelastic impact at increasing energies should correspond to these excitations' losses, leading to light emission, and not to ionization, the ionization potential being higher. It was found in complete conformity with Bohr's theory, the first inelastic impacts of electrons with atoms or molecules at lower energies, in general gives rise to the emission of light of the first line of a series of these atoms and that as the electrons' energies increase, the separate higher lines of appropriate frequency appears as the energy reaches a proper value.

At an appropriate energy of the impacting electron, the atomic or molecular electrons are completely removed from the atoms or molecules leaving behind the positively ionized atomic ions or molecular ions.

When an electron possesses more than an ionizing amount of energy, any superfluous energy which it has after causing ionization is distributed between itself and some electrons removed from the atoms or molecules.

A single electron of appropriately high energy can liberate as many as 4 to 5 electrons at once from an atom as in the outer electrons of mercury. The work on dissociation and the mechanism of ionization in certain gases, such as the rare gases nitrogen and hydrogen, has shown the possibility of the simultaneous excitation and ionization of the same atom by a single electron impact of appropriate energy.

It is found with few exceptions that the larger the diameter of the atoms and the larger the number of external electrons, the smaller the excitation and ionization potential will be.

The probability of resonance and ionization is greatest at the precise ionization or resonance potential and falls off exponentially from that value on, only to rise again as the next potential is reached. In any case, it can be definitely stated that the probability of ionization and excitation, calculated on the basis of the electron free path in a given gas, is at its maximum at the ionization or excitation potential; thereafter, the probability drops rapidly to lower values as the electron velocity increases.

We experimentally found the number of ionizing collisions per meter made by an electron is approximately proportional to the excess of energy of the electron above the ionizing energy, i.e., $n = C(E - E_1)$, where E is the actual energy of the electrons, E_1 is the ionization potential of the gas, and C is a constant.

The energy an electron must have before it can produce an appreciable amount of ionization is always considerably larger than the ionization potential. The ionization potential determines only the energy at which ionization by collision starts. The greatest amount of ionization occurs when the electron has an energy five to ten times that given by the ionizing potential.

The minimum value of the potential at which definitely inelastic impacts of electrons with atoms or molecules set in and where electrons lose all or a large fraction of their energy at a single impact, is called the critical potential. It was first observed that the conductivity of gas was increased at the critical potential, or first ionization potential; that is, when an electron acquired an energy of equivalent voltage, V_0, characteristic of a given gas, it was able on impact to remove an electron from an atom which has an ionization potential of V_0. This is expressed in electron volts. The second ionization potential corresponds to the work required to remove a second electron from the atom or molecule when the atom or molecule has already been ionized.

Unless ionization occurs, atoms and molecules can absorb only discrete amounts of energy. This energy has the effect of moving the most loosely held orbital electron or electrons in an atom to some larger orbit. Thus at normal temperature (80 degrees Farenheit), the average kinetic energy of the gas molecules corresponds to less than 0.04 e.v.; that is, the fraction of the total number of molecules which has energies greater than that necessary for ionization of the gas is extremely small.

Because the mass of the electron is so very small, the energy it loses in an elastic collision is only a small fraction of its total kinetic energy approximately equal to the ratio of the mass of the electron to the mass of the molecule even when the average energy of the electrons is only 20% or 30% larger than the average energy of the molecules. Thus with a comparatively small electric field, if the electrons can make a sufficient number of collisions the average energy of the electrons can become many times that of the molecules.

If the electric intensity were only 100 volts/meter, the average energy of the electrons in oxygen would be about 4.5 e.v. at 1 mm. Hg. This corresponds to a temperature of 35,000° K. Under conditions such as these the actual kinetic energy of a fairly large fraction of the electrons would thus be larger than the ionization energy of the molecules and ionization could occur owing to the collisions of the electrons with neutral molecules.

The forces acting between an electron and a molecule vary much more than the inverse, square law coulomb. This is represented by the following equation:

$$f = \frac{(D-1)\,Q^2}{10\pi^2\, r}$$

This assumes spherical elastic molecules and ions, these attract each other at a distance *r* with a force *f*. Thus, an electron can approach very close to a neutral molecule before experiencing any force due to the nuclei and the orbital electrons. If then, the electron has made a series of collisions and its energy or velocity is sufficiently low it can attach itself easily to the neutral molecule to form a negative ion. However, at too low pressures, the electron mean free path is increased and collisions are few thus insufficient to bring about the formation of negative molecular ions.

Electrons of low velocity approaching an ionized atom or atomic ion or molecular ion, must be able to interchange velocities so that while one electron neutralizes the ionized atom, the other electron escapes with the total energy resulting from the process. Another example would be a free, slow moving electron approaching an excited atom, the energy of excitation is given to the slow electron while the excited electron returns to its normal orbit without radiation or to some intermediate orbit with radiation of lower frequency. A classical example for illustration is the irradiation of mercury vapor by the line 2537 A°; the mercury vapor becomes activated and is then in a metastable state; if these atoms collide with thallium atoms while in this state, a thallium electron will be raised to a higher level so that it emits the green thallium line. The difference between the energy of the 2537 A° line and the low energy thallium line is converted into kinetic energy of the separating mercury and thallium atoms after impact. If the activated mercury atom strikes molecules in the excited state, the energy is converted into the work, a dissociation of these molecules into atoms or molecular hydrogen fragments. If the excited mercury atom collides with an atom of lower ionizing energy, this may remove an electron from the neutral atom ionizing it, and itself returns to the normal state.

Methods of electronic excitation:

A. Processes in the gas itself
   1. Rapidly moving electrons and beta particles from radioactive changes
   2. Rapidly moving positive charges, protons and alpha particles
   3. Rapidly moving positive ions in high electrostatic fields
   4. Photoelectric ionization by ulraviolet light, x-ray and other rays beyond the light rays as coming from the cosmos
   5. Through chemical reactions in the gas, e.g. oxidation of $NO \rightarrow NO_2$, $P \rightarrow P_2O_3$, P_2O_5, etc.
   6. Possibly ionization and excitation of the gas may be caused by temperature alone without action of the walls, or by impact of rapidly moving neutral atoms or molecules from other sources, for example neutralized alpha rays.

B. Processes due to solid or liquid surfaces in contact with gas
   1. Bombardment of metal by fast electrons, alpha particles, positive ions or recoil atoms giving secondary ions
   2. Action of metastable atoms on solid surfaces giving secondary ions
   3. Action of gamma rays, x-rays and light on solid or liquid surfaces giving electrons photoelectric effect
   4. Incandescent metals in general emitting large numbers of electrons called thermo-ionic emissions
   5. Incandescent surfaces having salts, phosphates, oxides, chlorides, or complex metal salts of mixed composition, at lower temperatures give positive ions such as sodium ions, potassium ions, barium ions, etc.
   6. Chemical reactions at surfaces, oxidation of moist phosphorus, potassium, sodium, give ions of both signs, but more negative ions.
   7. The atomizing of liquids into minute droplets by high velocity air

currents tangential to the surface causes the smallest droplets to become negatively charged. If water is used which contains ions, the larger droplets have ions in them and are predominately positive. The source of conductivity of the gas is the source of the charges on rain clouds and thunderstorms.

8. Frictional effects between solid particles suspended in gases, as typified in electrons in sandstorms
9. Ionization of a gas can occur when the average energy of the molecules becomes so great that the energy transferred in a collision between two neutral molecules is sufficient to ionize one of them.
10. The collision of free electrons with neutral atoms or molecules — ionization by collision between molecules and electrons in thermal ionization may involve several processes:
    a. The electrons may ionize directly in colliding with a neutral molecule.
    b. The electron may excite a molecule and the subsequent electron may ionize it.
    c. An electron might excite a neutral atom which subsequently in returning to its normal state would give off radiation and cause photo-ionization either in the gas or the walls of the discharge.
11. The electric field is then one of the most important ionization agents.

An ion moving under the influences of the forces of the electric field is given by the following equations: F = XAned, where F is the force, X is the field strength, A is the base area, *n* is the number of ions per cubic centimeter, *e* is the charge, and *d* is the length parallel to the field.

This force acts on each ion between its 10^9 collisions per second with molecules to give it momentum in the field direction. At each of the 10^9 impacts, some of this momentum is yielded to the neutral molecules with which the ion collides. As a result, the molecules are set into a motion along X.

When a molecule captures a charge to become a molecular ion, it moves in the electric field according to the following equation:

$K = \frac{e}{6\pi} - a\left(\frac{1 - AL}{a}\right)$, where *e* is the charge, K is the coefficient of viscosity, *a* is the radius of the particles, A is a constant of 0.874, and L is the mean free path.

The distance a molecule moves between impacts is called a free path, the mean free path is dependent on the velocities of the ions.

Now an ion that has fallen through a potential difference of one volt will have a kinetic energy of 1.6×10^{19} joule. That is, $qE = 1 \times 1.6 \times 10^{19} = 1.6 \times 10^{19}$ joule. According to the equation, $qE = \frac{1}{2} mv^2$, we find that the velocity of the particle after falling through a difference in potential of E volts is entirely independent of the length of the path the particle has traversed and also entirely independent of the form or shape of the electric field. The electric field intensity may be distorted in any way we please. That is, it may be at a high at one point, low at another, but still if the total difference in potential is E volts the velocity of the particle will be:

$v = \sqrt{\frac{2qE}{m}}$. For example, an electron with the kinetic energy of one electron volt has a velocity of $v = 5.93 \times 10^5$ M/sec.

A hydrogen ion having a kinetic energy of one electron volt would have a velocity of 10.85×10^3 M/sec., or nearly 6.1 miles per second. Hence the mobility of charged particles, positively charged molecular ions and negatively charged molecular ions or atomic ions in the electric field may be controlled by simply controlling the difference of potential through which the particles are falling. Further, since the molecules are to react with one another, the rate of reaction and the intensity of said reaction may be controlled at will.

A specific case in which the electric field performs the double function of molecular excitation and the creation of intermolecular and atomic ions is being given by the system used by the inventor.

It is a system utilizing the principles of the wire corona with a concentric cylinder at different pressures. The system is modified in conformity to the concept that chemical reactions must take place when the oppositely charged molecular ions from an appropriate activated catalyst are accelerated against one another in the wire corona. It consists of a cylinder made of a suitable catalyst from which positive atomic ions are emitted. The reactants (gases) streaming through the chamber parallel to the length of the wire attain the polarity of the negative molecular ions by the high electric field close to the wire. As these negative molecular ions are accelerated at the right angles to the wire in the direction of the electric field toward the positively charged catalyst cylinder, they are met by an avalanche of onrushing atomic ions from the catalyst. A certain amount of reaction takes place in that instant, 10^{-8} sec. However, some of the negative molecular ions outside the mean free path of the positive atomic ions are free to rush headlong toward the positive cylindrical field where they are neutralized, and instantly given a positive charge by the avalanche of outrushing positive ions. These positive molecular ions are accelerated back into the field and collide against the negative molecular ions coming from the direction of the negative electrode corona. This melee continues until the reaction has come to a point where the individual participants are either all gone or the mixture is outside of the electric field: backrush oscillations.

The Moray apparatus combined with other equipment, consists of a combination of specially constructed tubes which we will refer to as valves, "pressure transmitters," interceptors and oscillators. The valves are not rectifiers in the sense that they operate as radio valves in changing AC or HF to DC. They have an actual valve action in stopping the "flow" of energy which may be thought of as oscillatory action similar to the waves of the sea, without rectification, from returning to the outer circuit, much as a retaining wall could stop the waves of the sea from returning. The other modalities and "tubes" of the device are equally unique in their performance. Although no new laws of energy are being advanced or claimed as having been discovered, the application in the method of utilization of the energy throughout space is unique in that "generation" is accomplished by oscillatory utilization rather than by the conventional prime mover. These detector tubes have a synchronized pull with the specially developed oscillators of high faradic capacity and provide a means though which oscillating energy may pass to specially constructed valve oscillators whose relation to the first stage valve is such as to permit oscillations to come in from but not to return to the outer circuit with an automatic variable relation to the oscillations from the universe, and capable of setting up within their circuits initial oscillations which coincide with the oscillations of the universe.

Special provision is provided to stop RE tubes from becoming blocked in their dissipation of the charges created by the oscillations that continually accumulate based on the oscillatory capacity backrush effect common to capacitors and herein applied in vacuum tubes. This action of these devices has the effect of enlarging and prolonging the time of charge and discharge of the capacitors and the capacity energy in the circuit to an appreciable interval in perfect harmony with the natural energy wave through the interceptor's valves and oscillators in the circuit which set up in the circuit electrical pulsations corresponding to the energy waves captured by the interpretor and again kept from returning to the second outer circuit by "multi-walled" valves. The final tubes act as energy pressure transmitters with a means to prevent "shunting" condensation by a special form of "getter." This stops condensation accumulating at the base of the tubes which would block their ionic action.

One must "split" the energy discharge band into lines of variation (call this what you will), lines of energy or lines of light beyond the "light rays." The oscillations, therefore, do not become simple oscillations but through the action of the universe set up an energy flow which might be referred to as the assertion of inertia. When inertia sets in, the action will continue because of the oscillations of the cosmos, otherwise one would have a complete dissipation of energy and no oscillations. The oscillations will vibrate during the same period of time regardless of the potential, but the rate of vibration of the device depends on the "capacity" of its modalities, i.e., condensers, etc.

Let us go from the known to the unknown. We have referred to a form of ionic action in place of the common electronic liberation accomplished in radio tubes. It is an accepted fact that when various substances are bombarded with alpha particles they are found to give off electrons. This is the principle involved in various vacuum tubes. Thomson gave a similar action of liberation the name of "delta" rays. These delta rays of electrons are thought by some to originate in a type of ionization which results in "thermions," given off when the alpha particles strike the bombarded substance. It is, therefore, possible that some "particle" from the cosmos, with greater penetrating power than alpha particles, could penetrate quartz or various substances and set up a decided ionic action. Much could possibly be learned from a glorified "monochromator."

Just as sodium, potassium, cesium, rubidium, barium, and strontium react to visible light or wavelengths within a certain range, might not certain other substances react to oscillations from the cosmos or artifically produced radiations?

The universe is analogous to a radio transmitting station. It is continually emitting energy, only of a greater range of wavelengths. Wavelengths and frequencies are truly the answer to all vibrant worlds of living things. The steller laboratories provide environments as yet unproducible by man, or perhaps only unrecognized. The terms cosmic energy and Radiant Energy (as used by Dr. Moray) can be thought of as synonymous terms referring to frequencies of unknown and undefined limits. Parameters of these arbitrarily named ranges of the spectrum have not been defined so that the terms can be recognized as specific localities in the wavelength scale. Energy must be absorbed to be utilized. Absorption converts energy into heat, chemical energy, mechanical energy, electrical energy and perhaps into forms unknown at the present time. "Vibrant with life" is more than a poetical phrase.

In its simplest form an ion consists of a molecule of air that has either one or more, or one less than the quota of electrons for the electrically neutral molecule. The former is a negatively charged, the latter a positively charged ion. Positive ions are attracted toward negatively charged bodies while negative ions drift away from them. This process goes on, and the medium loses its electrical charge at a rate that is proportional to the abundance of ions and to the velocity at which the ions move towards the attracting medium. The velocity is less or greater if the ion is not the simple type but consists of a variable aggregate of molecules. Much could be written on this but the above should suffice for the present purpose. Let us go on from and not stop at the ultraviolet light theory of Aurora.

In some electronic tubes the electrons are not emitted directly from the filament but from an indirect cathode which does not enter into the direct electrical function of the tube. Does this teach us anything about an indirect generation of ions and ionic action? Could the opposite deflection of alpha and beta rays and the undeflected course of the gamma rays teach us anything about cosmic energy or Radiant Energy valves and oscillators?

The more perfect the ionic action the greater the collecting voltage, the greater the ionic gain of energy will be between collisions and the greater the amount of kinetic energy conserved. The collision will be "perfectly resilient."

The little we know about "space" and what it contains, or may contain, is so limited that we are forced to acknowledge that anything is possible beyond our experience. The actual proportion of matter to space contained in the celestial bodies is a very small part of the whole. There may be many pervading energies or matter that are more important than those we have detected. The very fact that we have heretofore been unable to detect them may make them all the more important in the cycles of our life and the things we know.

There is a multiplicity of phenomena which occurs at the same time in gas conduction; the known laws are largely empirical and approximate. Ohm's Law is valid in only a few limited cases, conductivity changes markedly with the variety of gas and the gas pressure. In the detector and intercepter circuits it becomes more important to maintain oscillatory action than frequency stability.

The relationship and combination of natural vibration and forced vibration is particularly important. It must be remembered that a point of resonance will be reached at some frequency, and the charge will reach a magnitude which depends upon the impressed force which is, in the case of the universe, immense. This is the effect of pure resonance.

Up to this point, science has been unable to produce pure resonance, but science is the accomplishment of that which was once thought to be unattainable. Science also, at one time, said it was impossible to transmit the human voice over a copper wire. The results are obtained through effort. Science also said it was impossible for a heavier-than-air device to fly.

Resonance has been obtained where a return wire in an electric circuit is not necessary. Pure resonance offers many dangers to the inexperienced investigator. The walls of Jericho are an example of the power of pure mechanical resonance. The most commonly known example of this phenomenon is the breaking of a glass by the power of the human voice. Marching feet, running dogs, and some harmonics set up vibrations which are dangerous for bridges and other mechanical structures. Every substance has a natural mechanical vibratory resonance and a point of dissociation. This is also true in electrical or energy resonance.

The theory is that in perfect resonance the oscillations will become more and more vigorous until the vibration or oscillation will go on forever if properly "fed", or the dissociation of the matter will result. However, while this will be true in theory, it is only partly true. In fact, if one could obtain resonance with the universe with its multiplicity of vibrations which correspond to the natural vibration and pure resonance of the different substances (each substance has a different pure sonic, pure vibration, pure resonance) buildings might be made to fall, glass to break, matter to explode. The writer has, as Tesla did, come very close to pure resonance with some substances, but achieving resonance with one substance is not being in resonance with all substances. If we could find the proper mechanical or energy resonance of certain kinds of matter, the constructive results would far outweigh the destructive. Gasoline and fire each have their place which must be honored and feared as well as appreciated. Mankind already possesses the technology and the capability to totally destroy the earth, and even though the control of pure resonance would add to this capability, it would contribute immeasurably to the good of mankind.

Put together in pure energy resonance certain energy responding apparatus which synchronize with the resonance of certain vibrations of the universe, and what do you have? Useable energy from the universe. This energy may come to the planets as oscillations similar to the oscillations and tides of the sea. The RE tubes received this energy in surges which may last only a few microseconds but the pressure and the current in those surges are so strong that sufficient energy is delivered to the equipment in resonance to be useable in multiples of flashes and in a magnitude which competes with the light of day. Remember resonance and pressure can do a lot to amplify energy. Also remember that the vibrations going out from the sources in the universe must also return to their sources. Nothing is lost. There is only a lowering of potential like water over the wheel.

The RE tubes present no new laws of physics. They simply expand the application of known laws thereby obtaining results not at first deemed possible. This is the history of science. RE tubes possess greater ability to obtain "saturation" and thus charge the accompanying capacitors or condensers at a more steady rate. When a certain voltage is reached, ionization occurs in the gases of the discharged tube and causes the condensers of the valve circuit to discharge into other condensers of the valve circuit to discharge into other condensers of the oscillators and the other modalities of the circuit.

When ionization in the preceding tubes is no longer possible because of the reduced voltage, the process starts all over again. The first valve passes vibrations of energy into an oscillatory circuit; ionization sets in, a discharge occurs, and energy passes through another valve into other oscillators. The process is repeated from the first stage, on to the second stage, on to the third and so on, much like a bucket brigade. That is why I asked years ago, "Cannot a steady flow of water be obtained from the waves of the sea or energy from the vibrations of the cosmos?"

Many phenomena, especially those occurring in certain frequencies, are still unexplained and there are numerous places where the classic theory and observed facts do not agree.

When a vibration of any kind strikes a boundary between two media of different vibratory impedances at an angle of less than 90 degrees, a transformation of the vibratory rate may be changed into another vibratory rate. The RE device therefore will continue to capture energy by resonance, or call it what you will, as long as the "keep alive" vibration of the cosmos continues to oscillate the various stages of valves and oscillators in the circuit. Simple, is it not? Just a case of the trapping of energy which is everywhere present in the primary circuit and causing it to oscillate through the secondary circuits through a blocked circuit of no return.

Our experiments have proved that there is an energy which exists in the universe which, by proper development of equipment, can be made available for commercial use.

One may say all energy comes from the sun. Can one prove the sun is the foundation of all energy? Or is the sun re-transmitter of energy? That light is an electrical phenomenon has been amply proved. The atoms in those distant stellar crucibles have moving electrons which are emitting electromagnetic waves of many lengths and many frequencies, which can be tuned to certain ranges of wavelengths. Our eyes and other senses repond to some of these frequencies, but there are many beyond those which we loosely term light. The photographic plate records some of these and also invisible radiations of shorter wavelengths or higher frequencies known as ultraviolet radiations. There are radiations measured by their heating effect which possess longer wavelengths or lower frequencies which we call infrared radiations. There are electromagnetic radiations of shorter wavelengths. These are generally known as roentgen rays. There are rays of still shorter wavelengths. These are of unlimited power, born and put into locomotion from the very source or foundation of energy. What is a man to do with such a picture of the universe other than let his interest and admiration grow? But will we ever get to the final foundation? Anything is possible beyond our experiences. And as Tennyson said:

"One God, one law, one element
And one far-off divine event
To which the whole creation moves."

Energy is vibration. Matter is a vibrating medium. All substances are really combinations of one primordial substance, i.e., vibrations. Electrons in motion constitute an electric current. What electricity is to matter, so is electric force to common mechanical force and electrical inertia to mechanical inertia. By inertia, we mean the ratio of force to acceleration.

Here on earth we have many receiving stations which are tuned to certain ranges of wavelengths.

Energy was defined in 1892* as a condition of matter, in virtue of which, any definite portion may effect changes in any other definite portion. Later discoveries have since confirmed this. Energy then, is a state of matter, or rather, the result of a particular state or condition in which matter may be, when any observed phase of energy appears.

Cosmic rays or cosmic vibrations and matter may appear to consist of entirely different things, but the fact is, these two subjects are actually joined together. Those highly penetrating rays which we call cosmic rays originate somewhere in the remote spaces of the universe and bombard our earth with continuous vigor, day and night, year after year.

We must not think of cosmic rays, infrared rays, ultraviolet, x-ray, gamma or any other ray or particles as simple in character. None of them in any sense of the word consists of a simple frequency of ray. Ultraviolet rays, x-rays, gamma rays and so on all consist of various wavelengths of frequencies. That is, all ultraviolet rays are not of the same frequency, and all x-rays are not of the same frequency. All are exceedingly complex in their make-up. This complexity is what has given them such an important place in the study of the physical sciences.

The great study of matter and cosmic rays, so closely tied together, will open up greater and greater fields of science for the practical good of the human race.

Experiments have also satisfied us that gravity is akin to an "electrical" oscillation, so well balanced that we might, for the lack of a better name, almost call it a "watt-less energy" until some force is exerted to oppose its potential, and then gravity opposes this force. This means gravity can be controlled and unlimited advancement made in air navigation.

Matter is capable of dissociation fitted to lead it into forms in which it loses all its material qualities. Among the most important dissociations Dr. Gustave LeBon noted in the 1890's is the emission by all matter of particles endowed with immense speed, capable of making the air a conductor of electricity, of passing through obstacles, and of being thrown out of their course by a magnetic field.* None of the forces then known were able to produce such effects, particularly the emission of particles with speeds approaching or equal to that of light; it was evident that science then found itself in the presence of absolutely

***The Evolution of Matter,** *by Gustave LeBon*

inexplicable facts. Several theories were put forth in explanation of them. One only — that of the dissociation of atoms, which Dr. Lebon advanced at the commencement of these researches — has resisted all criticism and on this account has now been almost universally adopted.

It is nearly seventy years since Dr. LeBon proved by experiment that the phenomena observed in substances termed radioactive — such as uranium — could be observed in all substances in nature, and could only be explained by the dissociation of their atoms.

The action of light on any substance, a lighted lamp, chemical reactions of very different kinds, an electric discharge, etc., causes these effluxes to appear. Substances termed radio-active such as uranium or radium simply present in a high degree phenomena which all matter possesses to some extent. When Dr. LeBon formulated for the first time this generalization, though it was supported by very precise experiments, it attracted little attention. In the whole world one physicist, the learned Prof. deHeen* alone grasped its importance and adopted it after having verified its perfect correctness. But the experiments being too convincing to permit of long challenge, the doctrine of the universal dissociation of matter at last triumphed. The atmosphere was then cleared, and physicists no longer denied that this dissociation of matter — this radioactivity as it is now called — is a universal phenomenon widely spread throughout the universe, and, as Prof. J.J. Thomson demonstrated, exists in most substances — water, sand, granite, clay, bricks, etc.

What becomes of matter when it dissociates? Can it be supposed that when atoms dissociate they only divide into smaller parts and thus form a simple dust of atoms? We shall see that nothing of the sort takes place, and that matter which dissociates de-materializes itself by passing through successive phases which gradually deprive it of its material qualities until it finally returns to the rate of vibrations whence it seems to have issued as energy.

When the fact is once recognized that atoms can dissociate and reappear as energy and then pass from energy back to matter, then the question arises where they obtain the immense quantity of energy necessary to launch into space as particles possessing a speed equal to the speed of light or **greater.**

The explanation in reality is simple enough, since it is verified that far from being an inert thing only capable of giving up the energy artifically supplied to it, matter is an enormous reservoir of energy, intra-cosmic energy.

Such a doctrine assails too many accepted fundamental scientific principles to be at once admitted. Accustomed to regard the rights and rigid principles of thermodynamics as absolute truths, and persuaded that an isolated material system could possess no other energy than that supplied from without, a majority of physicists long persisted in seeking outside of it the sources of the energy manifested during the dissociation (not disintegration) of matter.

The reality of this form of energy is proven by experimental facts. Cosmic energy is the most powerful of known forces and is the origin of most others, including intra-atomic energy.

* *ibid.*

Matter, formerly regarded as inert and only able to give back the energy originally supplied to it, is, on the other hand, a colossal reservoir of energy, intra-atomic and intra-cosmic energy, which can be expended without borrowing anything from without.

It is from the intrasonic energy manifested during the dissociation of matter that most of the forces in the universe are derived, notably electricity and solar heat.

Force and matter are two different forms of one and the same thing. Matter represents a stable form of intra-atomic energy; heat, light, electricity, etc., represent unstable forms of it (cosmic energy).

In the dissociation of atoms, the stable form of energy termed **matter** is simply changed into those unstable forms known by the names of **electricity, light, heat,** etc.

For the examination of these several propositions let us, as a basis of presentation, take them as proven, and seek at once the changes they bring about in a general conception of the mechanism of the universe.

The problem of the nature of matter and of force is one of those which has most exercised the sagacity of scientists. Its complete solution has escaped us because it really implies knowledge, still inaccessible, of the first cause of things. Scientific theories heretofore set forth have not allowed us to completely solve this great question. They lead, however, to a conception of matter and energy far different from that commonly in use at the present day. Cosmic energy will lead to some serious revision of current theory.

We can arrive at the conclusion that matter is an immense reservoir of energy solely constituted by a system of vibrating atoms maintained in equilibrium by the rotations, attractions and repulsions of matter's component parts. The material properties of bodies such as weight, form, and apparent permanence arise from this equilibrium. Matter also represents movement, but the movements of its component elements are confined within a very restricted space. This conception leads us to view matter as a variety of energy. To the known forms of energy — heat and light — there must be added another: energy from the cosmos which embraces but surpasses intra-atomic energy. This cosmic energy is characterized by its collossal greatness and its unlimited accumulation within everything in the universe.

It follows from the preceding statements that by postulating the dissociation of matter one is simply adding to the variety of energy which fills all space in different forms. For example, we call two forms of energy **electricity** and **light,** but in reality they are radiations or forms of vibrations.

We will endeavor to give an account of the forms under which this energy of the universe may be condensed within the atom and the cosmos, but the existence of the fact itself has a far greater importance than the theories of simple atomic energy it gives rise to. Without pretending to give the definition of energy so vainly sought for, we will content ourselves with stating that all phenomenality is nothing but a transformation of equilibrium. When the transformations of equilibrium are rapid, we call them electricity, heat, light, all forms of vibration. When the changes of equilibrium are slower, we call them matter. To go beyond this we must wander into the region of hypothesis and admit that the elements of the aggregate are represented by forces in equilibrium and are constituted by vortices formed in the midst of the universe. These vortices possess an individuality, supposed by some to be ephemeral but proved by the evolution of matter and energy to be eternal. The individuality disappears and the vortex dissolves as soon as the forces which maintain their existence cease to act; but others form elsewhere, i.e., eternal creation, eternal evolution, eternal energy and matter.

The equilibriums of these elements of which the aggregate constitutes an atom, may be compared to those which keep the planets in their orbits. As soon as they are disturbed, considerable energies manifest themselves, as they would were the earth or any other planet to be jolted in its course.

Such disturbances in planetary systems may be realized as they are in very radioactive bodies when, for diverse reasons, they have reached a certain degree of instability, or artificiality. This also occurs in ordinary bodies when brought under the influence of various excitants.

These excitants act in such cases like the detonator of an explosive; they free quantities of energy greatly in excess of the very slight cause of their liberation. And, as the energy condensed in the universe is immense, an extremely slight loss in matter results in the creation of an enormous quantity of energy.

From this standpoint, we may say of the various forms of energy such as heat, electricity, and light that all vibrations represent the last stages of matter before it returns into the cosmos from whence it came.

If, extending these ideas, we wish to apply them to the differences presented by the various simple bodies studied, we should say that one simple body only differs from another in rate of vibration. If we could deprive any element of a sufficent quantity of the energy it contains, we should succeed in completely transforming it.

As to the necessarily hypothetical origin of the energies condensed within the atom, we will seek for it in a phenomenon analogous to that invoked by astromers to explain the formation of the sun and the energies it stores. To their minds this formation is the necessary consequence of the condensation of the primitive nebula. If this theory is valid for the solar system, an analogous explanation is equally so for all matter.

Such a theory clears away the classical duality of matter and energy. They thus become aspects of the same thing. There is no separation between matter and energy, since matter is nothing more than a stable form of energy, continually changing from one to the other in the cosmos.

We can only understand things by fitting them into the common frame of our thoughts. The essence of energy being unknown, we are compelled to materialize it in order to reason about it. We thus arrive — but only for the purposes of demonstration — at the following definitions: Energy and matter represent entities of the same order; the various forms of energy — electricity, heat, light — are manifestations of matter in action and only differ in the nature and the stability of the equilibrium formed in the heart of the cosmos. It is through these manifestations that the universe is known to us.

The illustrious Faraday endeavored to clear away the duality existing between matter and energy. Others made the same attempt by pointing out that matter is only brought home to us by the intermediary of forces acting on our senses. But all arguments of this order were considered as having a purely metaphysical bearing. In Faraday's time it was argued that it had not been possible to transform matter into energy, and that this matter was necessary to animate the former. Scientific principles, considered assured, taught that nature, as a kind of inert reservoir, could create the liquid it holds. Everything seemed then to point out that nature and energy were irreducible things as independent of each other as weight is of color. It was therefore not without reason that they were taken as belonging to different worlds.

The transformation of matter into energy now having been demonstrated, it follows without question that energy can be transformed into matter.

The facts summed up in the preceding pages show that matter, in a set form, is not eternal, but as a rate of vibration or energy it is eternal; that it constitutes an enormous reservoir of vibrations; that it disappears by transforming itself into other forms of energy or matter before returning to its source. That it is, in reality, just another cycle in the process of creation.

It can therefore be said that if matter cannot be created, it cannot be destroyed; returning to its source, it begins a cycle again. The elements of a substance which is burned or annihilated by any other means are transformed by a change of vibration. They may lose every quality of matter — including the most fundamental of them all, weight — but that loss only shows that gravity lost its power over the elements because of a change in their vibration. The theoretical importance of these principles is considerable and self-evident.

It is easy to deprive matter of all its attributes, save one. Solidity, shape, color, chemical properties may disappear, but there remains a rate of vibration. The hardest body can be transformed into an invisible vapor. But, in spite of every one of these changes, the mass of the body, as measured by its weight, remains invariable, and by changing or restoring rates of vibration this mass can be made to reappear. Thus the one fixed point in the mobile ocean of phenomena is vibration. It enables the chemist, as well as the physicist, to follow matter through its perpetual transformations, and this is why matter remains something mobile and eternal.

The importance of permanence and the indestructibility of vibration which one recognizes throughout the changes in matter are the only characteristics by which this great concept can be grasped, and they necessarily become preponderant.

Discovery is rarely spontaneous; it only appears so because the difficulties and hesitations which most often surround its inception are generally unnoticed. The public troubles itself very little with the way in which inventions are made, but psychologists should certainly be interested in certain aspects of the problems of inventions. In fact, they will find therein valuable documents on the part played, even in laboratories, by suggestions and illusions, and finally on the preponderant influence of prestige which is a principal element of demonstration and much of the time supplants facts.

The researches detailed in this work tend to shake fundamental dogmas which are the bases of modern science. If the principle of the conservation of energy — which is simply a bold generalization of experiments made in very simple cases — likewise succumbs to the blows which are already attacking it, we must arrive at the conclusion that nothing in the world is eternal in a set form, but all is subject to change of rate of vibration. The great divinities of science could also be condemned to submit to that invariable cycle which rules all things.

On the ruins of former doctrines and after centuries or persistent effort, there sprang up two sovereign powers which seemed eternal — matter as the fundamental woof of things and energy to animate it. With the equations connecting them, modern science thought it could explain all phenomena. In its learned formulas all the secrets of the universe were enclosed. The divinities of old were replaced by ingenious systems of differential equations.

The discovery of the dissociation of matter by radiation has allowed us to penetrate into an unknown world ruled by new forces, where matter, losing its properties as we have known it, becomes a form which passes without difficulty through obstacles and possesses a whole series of unforeseen properties, the far-reaching effect of which we have yet to learn.

The discovery of the universal dissociation of matter is linked to that of intra-atomic energy and energy of the cosmos, by which science has succeeded in explaining some radioactive phenomena. The origin of intra-atomic energy is not difficult to elucidate if one supposes, as do the astronomers, that the condensation of our nebula suffices by itself to explain the constitution of our solar system. It is conceivable that an analogous condensation of the cosmos may have begotten the energies contained in the atom. The latter may be roughly compared to a sphere in which a non-liquifiable gas was compressed to the degree of thousands of atmospheres at the beginning of the world.

The reason this force (the most widespread and the mightiest of all those of nature) has remained so long unrecognized is that man lacked the reagents necessary for the proof of its existence. The atomic theory and edifice erected by science were so stable, so solidly fixed in our minds, that its dissociation and connection with the cosmos seemed extremely unlikely. Had it been otherwise, the world would long ago have utilized cosmic energy.*

The generality of the phenomenon of the dissociation of matter would have been noticed much sooner if a number of known facts had been closely examined. This is also true of the law of oscillations by vibrations of the universe. These facts were spread over very different chapters of physics. For example, the loss of an electric charge occasioned by ultraviolet light has long been known but not the thought connecting this fact with a cathode ray. More than 75 years ago Niepe de Sant-Victor saw that dark salts of uranium caused photographic impressions after several month's exposure, but as this phenomenon did not

**Let us here remind the reader that this theory was first given to the world by Henry Moray as early as 1914 and 1926 and was first published in printed book form in 1931.*

seem to be connected with any known fact, it was put to one side as the oscillations of the cosmos are now. For hundreds of years the gases of the oscillations or vibrations of the universe have gone unappreciated. The common link which connects them appeared clearly when we established that the dissociation of matter and the forms of energy which result from it are to be ranked among the most widely spread natural phenomena. Flames had been observed to discharge electrified bodies without anyone determining the exact cause of this phenomenon. The loss of electric charges through the influence of light had been known for many years, but it was regarded as a fact peculiar to a few metals with no suspicion of how general and important this law was.

All these phenomena and many others, such as electricity and solar heat, are very dissimilar in appearance, but are the consequences of the same process, the dissociation of matter.

But how is it that a demonstration so simple as that of the existence of cosmic Radiant Energy has not been made since the discovery of radioactivity, especially since the demonstration of the generality of radioactivity? This can only be explained by bearing in mind that it was contrary to all accepted principles that either matter or the energy of the universe could by itself produce energy. Scientific dogmas inspire the same superstitious fear and respect as did the gods of old, though at times they have been gods that failed.

The existence of condensation of energy within the atoms of the universe at first seemed to annoy us because it was outside the range of things formerly taught by experience. However, even leaving aside the facts revealed by radioactivity, it should be remarked that analogous concentrations are observable daily. Is it not strikingly evident, in fact, that enormous accumulations of electricity must exist in all substances of the universe if, by the electrolysis of water, it is found that one gramme of hydrogen possesses an electric charge of 96,000 coulombs? This suggests the degree of condensation in which electricity existed before its liberation in all the universe. Elementary treatises have long since pointed out that barely a twentieth part of the above quantity would suffice to charge a globe the size of the earth to a potential of many thousand volts. The best static machines of our laboratories hardly give forth 1/10,000 of a coulomb per second. Consequently, they would have to work without pause for more than thirty years to produce the quantity of electricity contained within the atoms of one gramme of hydrogen.

As electricity exists in a state of considerable concentration in all matter, it is evident that the atoms of the universe should have been regarded many years ago as a veritable capacitor of energy. We should have recognized that the quantity of this energy must be enormous; it was only necessary to appreciate the magnitude of the attractions and repulsions which are produced by electric charges before us. It is curious to note that several physicists years ago touched the fringe of this question without perceiving its consequences. For example, Cornu* pointed out that if it were possible to concentrate a charge of one coulomb on a very small sphere, and bring it within one centimeter on another

Compare this to Questioni Di Fisica Moderna, Vol. 1,* **Geometrodynamics, *V.A. Wheeler, Academic Press, 1962.*

sphere having a like charge of one coulomb, the repulsive force created would equal 9^{18} dynes, or about 9 billion kilogrammes.

What about the stars and planets of the universe? We have said that by the dissociation of water we can obtain from one gramme of hydrogen an electric charge of 96,000 coulombs. This energy would be enough (and this is exactly the hypothesis enunciated by J.J. Thomson) to dispose the electric particles at suitable distances within the universe to obtain through their attractions, repulsions, and rotations, extremely powerful energies in an extremely small space. The difficulty is not, therefore, in conceiving that a great deal of energy could remain within any atom (not just a radioactive one), but that nature is supplying us from the universe useable energy if we but reach out to harness it. It is surprising that a notion so evident was not recognized long ago.

Calculations of radioactive energy has been made within those limits of speed at which experiments show that the inertia of matter does not greatly vary.

Science formerly established a complete separation between matter and energy. The classic ideas on this decision are shown in the following passage of a work by Prof. Janet:

> "The world we live in was thought, in reality, a double world; or rather it was composed of two distinct worlds — one the world of matter, the other the world of energy. Copper, ion, and coal are forms of matter, mechanical labor and heat are formes of energy. These two worlds are each ruled by one and the same law. Matter can neither be created nor destroyed. Energy can neither be created nor destroyed, therefore we can rightly observe that as energy and matter are one in different stages, we have but one world in the evolution of matter and energy.
>
> "Matter and energy can assume various forms without matter transforming itself into energy or energy into matter — we can no more conceive energy without matter than we can conceive matter without energy.
>
> " 'It is utterly impossible,' Lord Kelvin wrote, 'that the heat produced can proceed from the stored energy of radium. It therefore seems to me absolutely certain that if the emission of heat continues at the same rate, this heat must be supplied from outside.' "*

Lord Kelvin fell back on the commonplace hypothesis formed at the outset on the origin of the energy of radioactive bodies, which was attributable, it was then believed, to some mysterious force from the ambient medium. The supposition that matter was entirely unable to create energy and could only give back what had been supplied to it had no experimental support. The fundamental principles of thermodynamics which Lord Kelvin had helped so much to found, tell us, in fact, that a material system isolated from all external action cannot spontaneously generate energy. Experiment is superior to theory, and when once it has spoken those scientific laws which appeared to be the most stable will be condemned to join in oblivion the used-up, worn-out dogmas and doctrines of the past.

Even in this so-called atomic and space age, it would be desirable to have a

*"*Lecons d'electricite,*" *Janet, 2nd edition, p. 2 and 5*

theory to explain the observed facts and to enable science to know whether the energy is borrowed from the atom itself or from external sources from the universe.

Many physicists, like Lord Kelvin, held to the end the old principles; that is why the phenomena of radioactivity, especially the spontaneous emission of particles animated with great speed and the rise in temperature during radioactivity, seemed to them utterly inexplicable and constituted a scientific enigma, as M. Mascart said. The enigma, however, is simply explained now. Today, there is another enigma in the use of cosmic power for commercial purposes.

One can hope, however, that ideas so opposed to classical dogmas as oscillatory cosmic energy and the transformation of matter into energy and energy into matter will soon be widely accepted.

The fact is that the scientific ideas which rule the minds of men at various epochs have all the solidity of religious dogmas. Very slow to be established, they are very slow likewise to disappear. New scientific truths, although they have experience and reason as a basis, are only propagated by prestige; that is, when they are enunciated by those whose official position gives them prestige in the eyes of the scientific world. Truths of such great importance as Ohm's Law, which governs the whole of electricity, and the law of the conservation of energy, which governs all physics, were first received with indifference or contempt and remained without effect until the day when they were enunicated anew by individuals endowed with authority. Now we limit the conservation of energy without fully understanding what truths it embraces.

It is only by studying the history of sciences that one succeeds in understanding the genesis of beliefs and the laws governing their diffusion. We have just mentioned two discoveries which were among the most important of the 19th century, and which are now summarized in the laws. One can say that they should immediately have appealed to all minds because of their marvelous simplicity and their imposing grandeur. Now tradition so fences them in that we fail to see their magnitude. Not only did they strike no one when they emerged, but the most eminent scientists of that time did not concern themselves with them except to try and cover them with ridicule. Are we different today?

That the simple enunciation of such doctrines appealed to so few shows with what difficulty a new idea is accepted when it does not fit in with contemporary dogmas. Prestige, we repeat, and to a slight extent experience, are the ordinary foundations of our convictions — scientific and otherwise. Even the most convincing experiments have never constituted an immediately demonstrable foundation when they have clashed with long established accepted ideas. We hate to discard the comfortable old dogmas even when they don't fit the new evidence. Galileo learned this when, having brought together all the philosophers of the celebrated University of Pisa, he sought to prove to them by experiment that, contrary to the accepted ideas, bodies of different weights fall with the same velocity. Galileo's demonstration was assuredly conclusive: letting a small leaden ball and a cannon shot of the same metal fall at the same moment from the top of a tower, he showed that both bodies reached the ground together. The professors contented themselves with appealing to the authority of Aristotle and would not change or modify their opinions. For years it has been the same with RE demonstrations, especially among those who never saw the device. Galileo lived a long time ago, but the degree of receptivity of minds for new things has not ostensibly increased.

When Ohm discovered the law which immortalized his name and on which the whole science of electrical measurement rests, he published it in a book filled with experiments so simple and so conclusive that they might have been understood by any child in elementary school. Not only did he fail to convince anyone, but the most influential scholars of his time treated him in such a way that he lost the position he occupied as a college professor, and, to avoid dying of starvation, was only too glad to take a situation in a college at 1200 francs per annum, where he remained for 6 years. Justice was rendered to him only at the close of his life. Robert Mayer, less fortunate, did not even obtain this belated satisfaction. When he discovered the most important of modern scientific laws — namely, that of all the examples of the conservation of energy, the vibration of the universe is the greatest example — he had great difficulty in finding a publisher who would consent to publish his findings; no scholar bestowed the least attention upon his pronouncement, no more in fact that on his subsequent publications, among which was the one on the mechanical equivalent of heat, published in 1850. After attempting suicide, Mayer went out of his mind, and remained for a long time after his death unknown, to such a degree that when Helmholtz re-made the same discovery, he was not aware of the work of Mayer. Helmholtz himself did not meet with any greater encouragement to begin with, and the most important of the scientific journals of that day, **The Annals Do Poggendorff,** *declined to print his celebrated memoir, "The Conservation of Energy," regarding it as a fanciful speculation unworthy of the attention of serious readers.*

Although matter was formerly considered inert and only capable of preserving and restoring the energy which had first been given to it, it was necessarily established that there existed within it forces (sometimes in considerable amounts) such as cohesion, which forces were independent of all exteral agents. Other forces, such as radiant heat and electricity, which also issued from matter, might be considered simple restitutions of an energy borrowed from the cosmos.

If that cohesion which makes a rigid block out of the dust of atoms, or if that affinity which creates chemical combinations, or if those osmotic attractions and repulsions which hold in dependency the most important phenomena of life, are visibly forces inherent to matter itself, then with the old ideas it was impossible to determine the source of this energy. The origin of these forces ceases to be mysterious when it is known that the cosmos is a colossal reservoir of energy that fills all space. Observation has long ago shown that any form of energy lends itself to a large number of transformations, and we can conceive how energy from the cosmos may be the source of all the molecular forces — cohesion, affinity, etc. — of matter. We are far from being acquainted with their character, but at least we see the source from which they spring: the universe where matter is "born" from energy.

Outside the forces plainly inherent to matter that we have considered, there are two — electricity and solar heat — whose origins remain unknown, and for which we can find an explanation in the theory of intra-atomic energy and inter-cosmic energy, the cradle of the intra-atomic energy.

When we study in detail the facts on which their theories are based, we find that electricity is one of the most constant manifestations of the dissociation of matter. Matter being nothing else but cosmic energy itself, it may be said that to dissociate matter is simply to liberate a little of the intra-atomic energy throughout the universe and to oblige it to take another form. Electricity is precisely one of these forms.

Throughout the years the role of electricity has constantly grown in importance. It is at the base of all chemical reactions; it is a universal force, and one must connect all other forms with it. That a force with the manifestation importance and universality of electricity should have remained unknown for thousands of years constitutes one of the most striking examples of apathy in the history of science, and is one of those facts we must always bear in mind when attempting to understand how we may be surrounded by other powerful forces without fully realizing their existence. Power from the cosmos, Radiant Energy, is another similar example.

For centuries all that was known about electricity could be reduced to this: certain resinous substances when rubbed together attract light bodies. Could not other bodies enjoy the same property? By extending the friction to larger surfaces might not more intense effects be produced? This was the one question to pose. However, ages passed before a mind arose penetrating enough to ask it — one inquisitive enough to test by experiment whether a body with a large surface when rubbed would exercise an action superior in energy to that produced by a small fragment of the same body. From this simple verification, which took centuries to accomplish, emerges the frictional electric machine and the phenomena it produces. Why not now let this phenomenon emerge from the oscillations of the universe and put into the hands of man a power which he thought the Gods alone possessed?

Electricity was once produced laboriously and was considered a very exceptional phenomenon. Now we find it everywhere and know that the simple contact of two heterogeneous bodies suffices to generate it. The difficulty now is not how to produce electricity, but how to prevent the production of unwanted electricity. The falling of a drop of water, the heating of a gaseous mass by the sun, the raising of the temperature of twisted wires, the burning of a match — any action capable of modifying the nature of a body generates electricity.

If all chemical reactions are electrical reactions, as is now proven to be the case — if the sun cannot change the temperature of a body without disengaging electricity, if a drop of water cannot fall without its manifestation — it is evident that electricity's role in all forms of life must be preponderant. This, in fact, is what the world is beginning to admit. Not a single change takes place in the cells of the body, no vital reaction is effected in the tissues, without the intervention of electricity. M. Berthelot showed the important roles of the electric tension to which plants are constantly subjected. The variations in the electric potential of the atmosphere are enormous, since they may oscillate between 600 and 800 volts in fine weather, and rise to 15,000 volts at the least fall of rain. This potential increases at the rate of from 20 to 30 volts per meter of height in clear weather or from 400 to 500 volts in rainy weather at the same elevation. "These figures," Berthelot said, "give an idea of the potential which exists either between the upper point of a rod of which the other extremity is earthed, or between the top of a plant or a tree, and the layer of air in which that point or that top is bathed." M. Berthelot also proved that the effluvia generated by these differences of tension can provoke numerous chemical reactions — the fixation of nitrogen on hydrates of carbon, the dissociation of carbonic acid into carbonic oxide and oxygen, etc. Why go back to these scientists of yesteryear? Only to show that much we are doing today does not originate with the atomic age! There were thinkers before our day. After having established the phenomenon of the general dissociation of matter, let us ask ourselves if the universal electricity, the origin of which remains unexplained, is not precisely the consequence of the universal displacement of matter. Experiments fully verify this hypothesis, and they prove that electricity is one of the most important forms of intra-atomic energy liberated by the displacement of matter. The various methods to obtain electricity, notably friction; only hasten the dissociation of matter. Now let us turn to vibratory oscillations of the universe.

CHAPTER 9
THE DISSOCIATION OF MATTER

As we study the dissociation of matter, the importance of this phenomenon proportionately increases. After recognizing that, electricity may be considered one of the manifestations of the vibration of matter.

To maintain that stars such as the sun can keep up their own temperature by the heat resulting from the dissociation of their component atoms, seems much like saying that a heated body is capable of maintaining its temperature without any contribution from outside.

Atomic reactions cool, simply because the rise in temperature produced during the dissociation of atoms producing the incandescence is far too slight to compensate for loss of heat by radiation. The substances which, like radium, most rapidly dissociate can hardly maintain their temperature at more than 3 degrees to 4 degrees centigrade above that of the ambient medium. Suppose, however, that the dissociation of any substance were only one thousand times more rapid than that of radium, then the quantity of energy it emitted could more than suffice to keep it in a state of incandescence.

The whole question therefore is whether, at the origin of things — that is to say, at the epoch when atoms were formed by condensations — those atoms possessed such a quantity of energy that by their slow dissociation they have been able ever since to maintain the stars in a state of vibration. This supposition is supported by various experiments. J.J. Thomson arrived at the conclusion that the energy now concentrated within the atoms is but an insignificant portion of that which they formerly contained and lost by radiation. Independently, and at an earlier date, Prof. Filippo Re arrived at the same conclusion. If, therefore, atoms formerly contained a quantity of energy far exceeding the still formidable amount they now possess, they may, by dissociation have expended during long accumulations of ages a part of the gigantic reserves of forces piled up within them at their source in the universe. They may have been able, and consequently may still be able, to maintain a very high rate of vibration, e.g. stars, like the sun and the heavenly bodies. In the course of time, however, the store of intra-atomic energy within the atoms of certain stars may at length be reduced, and their dissociation and rate of vibration may become slower and slower. Finally, they have acquired an increasing stability, have dissociated very slowly, and have become such as one observes them today in the shape of cooled stars like the earth and other planets.

If the theories thus formulated are correct, and the experiments of the Moray Radiant Energy discoveries so indicate, then intra-atomic energy manifested during the dematerialization of matter constitutes the fundamental element from which most other forces are derived. So it is not only electricity which is one of its manifestations, but also solar vibrations, the primary source of life and of the majority of the forces at our disposal. This study, which reveals to us matter in a totally new aspect, permits us to throw unforeseen light on the higher mechanics of the universe.

Interest is now being shown in spontaneously radioactive substances because they emit elements which other bodies only produce in much smaller quantity. By thus enlarging on this general phenomenon, we encourage its study in more detail, as well as the rate of vibration in all things.

Rutherford, who studied radioactive substances with great success and, with Curie, discovered facts concerning rays from them, led to the designation of the radiations as alpha, beta, and gamma. The alpha radiations are composed of positive ions, the beta radiations of electrons identical with those constituting the cathode rays, while the gamma radiations are similar to the x-rays. Did these pioneers build better than they realized in showing the way?

Alpha particles are formed of positive ions. They are deviated by an intense magnetic field, but in the opposite direction to the beta rays or particles. The radius of curvature of their deviation is 1000 times greater than that of the beta particles. They form 99% of the total radioactivity of radium. They render air a conductor of electricity. Their action on a photographic plate is much less than that of the beta rays, and their force of penetration very slight, since they are stopped by a sheet of paper. This weak power of penetration enables them to be easily differentiated from the other radiations to which paper is no obstacle. Of all the emission of radioactive bodies, it is the alpha particles especially which make the air a conductor of electricity, and it is the beta particles of rays which produce photographic impressions. When a radioactive body is enclosed in a glass tube, nearly all the alpha particles are stopped by the glass walls; but inside those glass walls great things take place because of the electrical conducting ability of the particles.

It is supposed, from various calculations, that the alpha particles must have a mass equal to or greater than that of the hydrogen atom and a like charge. Their speed, as calculated from the extent of their deviations by a magnetic field of given intensity, is one-tenth that of light. Their quantity varies according to the substance. For uranium and thorium, it is for one gramme, 70,000 emissions per second, and for radium a hundred thousand millions. This emission may last without interruption for more than 100 years.

The emission of alpha particles, positive ions, together with the production of the emanation, is the fundamental phenomenon of radioactivity. The emission of beta particles and that of the gamma rays, which together form hardly 1% of the total emission, should represent a further stage in the dissociation of radioactive atoms. How puny are these compared to the total energy in the cosmos, but how close the relationship.

On striking phosphorescent bodies the alpha particles render them luminous. It is this property which is the basis for the operation of the spinthariscope, an instrument which makes visible the permanent dissociation of matter. It consists of a screen of sulphide of zinc, above which is placed a small metal rod, the end of which has been dipped in a solution of chloride of radium. On examining the screen through a magnifying glass, there can be seen spurting out without cessation a shower of small sparks produced by the impact of the alpha particles, and the emission may last for centuries. This shows the extreme smallness of the particles coming from the disaggregation of atoms. This emission is visible because each particle is made apparent through the enormous degree of lateral perturbation produced by its shock on the sensitive surface, in the same way that raindrops falling into water produce ripples which exceed their diameter. One may, by using certain varieties of phosphorescent sulphide, succeed in making screens showing the phenomenon of dissociation, not only with salts of radium, but also with other substances. Herein is the door to greater discoveries.

The high speed of the alpha particles seems very difficult to explain. This speed is intelligible enough in the case of the beta rays which, being composed of atoms of pure electricity and having a very small inertia, can acquire a very high speed under the influence of limited forces; but with the alpha particles, whose dimensions would appear to be identical with those of the hydrogen atom, a velocity of 30,000 kilometers per second seems more difficult to explain. It could prove profitable if these experiments of Rutherford and his co-workers were taken up again.

These velocities may not be produced instantaneously; they are only comprehensible on the hypothesis that the particles of atoms can be compared to small planetary systems animated with enormous velocities. They would preserve their speed on leaving their orbits as does a stone launched from a sling. The invisible speed of rotation of the elements of matter would therefore be simply transformed into a speed of projection, perceptible only by proper instruments.

Beta rays are considered to be composed of electrons identical with those of the cathode rays. They are formed of negative electric particles freed from all matter. Their mass is similar to that of the cathode particles, the thousandth part at of the hydrogen atom. Their velocity varies between 33% and 96% of that of light.

They immediately render luminous by impact bodies capable of phosphorescence, even when separated from them by aluminum foil. The phosphorescene is very bright in platinocyanide of barium and some kinds of diamonds which are rather rare, but are capable of phosphorescence.

In addition to alpha and beta particles, the first charged with positive and the second with negative electricity, radioactive bodies emit an extremely slight proportion (less than 1%) of gamma rays, the properties of which are entirely analogous to x-rays but possess a higher power of penetration since they can penetrate several centimeters of steel. This property enables them to be easily distinguished from alpha and beta rays, which are stopped by a lead plate a few milimeters thick. (*Ed. note: Written by Moray in Europe in 1914 when these penetration values were considered valid.)*

One of the singular properties of alpha-beta-gamma emissions is that none of them can touch a gaseous or solid body without immediately causing, no doubt through the disturbance produced by their enormous velocity, a dissociation resulting in the production of secondary rays, which are similar in their properties to the primary rays, but less intense. These secondary radiations also expose photographic plates, render the air a conductor of electricity, and are deviated by a magnetic field. They are able to produce by their impact tertiary rays having the same properties. It is the secondary rays produced by the gamma rays which are the most active. A photographic impression through a metallic plate is sometimes intensified by the interposition of that plate, because the action of the secondary rays are then superimposed on the primary rays. Herein may lie a prototype of the action of the universe.

Emanations can be cheaply drawn from any highly radioactive substance: either by dissolving it in liquid placed in a receiver communicating with a closed tube and subjecting it to certain ray bombardment, or by bringing it to a red heat in a similar apparatus and bombarding it with rays. The emanation drawn into the tube renders it phosphorescent. It can be condensed in liquid air. This condensation is revealed by the localization of the phosphorescence.

At ordinary temperature, radioactive bodies in a solid state emit the emanation, but only a hundredth part of the quantity emitted in a state of solution and under bombardment of certain rays. By introducing sulphide of zinc into a bulb containing a solution of chloride of radium, the disengagement of the emanation renders the sulphide phosphorescent. Radium, when heated, loses the greater part of its activity because of the quantity of emanation it gives off, but it regains it again in about 20 days. The same loss occurs when a solution of this salt is heated to boiling.

When solid chloride of radium has been brought to a red heat, or a solution of it has been boiled for some time, or better still, it is subjected to special ray bombardment, it still preserves a quarter of its primary activity. This is then solely due to the alpha particles, as can be noted by the weak penetrating power of the rays emitted, which can no longer pass through a sheet of paper. It is only after a certain lapse of time that the appearance of the beta rays, capable of passing through metals, again takes place. The activity of the emanation is lost rather quickly. The rapidity of this loss varies according to the substance. That of actinium is destroyed in a few seconds, that of thorium in a few minutes, that of radium only at the end of three weeks, but it is reduced to one-half in four days.

Radium and thorium produce difference energy levels of emanations; that is, of dissociations which begin with the emission of the emanations. Five or six have been counted. The first engenders the second, and so on. They, no doubt, represent successive stages in the evolution of matter.

To these emanations are due three-fourths of the heat continually produced by radium. If radium be deprived of its emanation by heating, it gives out about a quarter of the heat it emitted before heating. Almost all of the rise in temperature is due to the alpha particles. If some emanation of radium is left for several days in a tube, one can observe the special lines of helium which were not there before.

Before drawing too many conclusions from this transformation, it must be first remarked that helium is a gas which accompanies all radioactive minerals. It was from these bodies that it was first obtained. This gas enters into no chemical combination; it will not liquify and can be kept for an indefinite time in the tube in which it is enclosed, a catalyst with an interesting effect.

This derivation from radium is a special helium since it appears to possess the property of spontaneously vanishing. Its only resemblance to ordinary helium seems to consist of the momentary presence of some spectral rays.

The emanation disintegration products of the radioactive bodies represent one of the intermediate substances. It is part material, since it can be condensed and dissolved in certain acids and recovered by evaporation. But it is only incomplete material, since it ends by entirely disappearing and transforming itself into electric particles and secondary particles. This transformation, which takes place even in a sealed glass tube, has been proved by experimentation.

Matter and energy! Where can one draw the line? The part played by various radiating substances in the phenomena of energy and life is a very predominant one. Most often it is the indirect reactions which reveal their existence in the phenomena of life and allow them to be isolated. All we know of them in their physiological reactions is that they lose their properties if deprived of the infinitely small quantities of mineral matter which they contain in a form that we suppose to be the ionic state.

Why go into these reactions? Because there is a parallel of phenomena which may to a degree help us to understand natural phenomena of the cosmos and of all life.

Vapor of water is found among the substances which become an essential part of the environment of certain reactions. Even in extremely small amounts it plays an important part in various reactions. Perfectly dry acetylene is without action on hydride of potassium, but in the presence of a trace of humidity the two bodies react one on the other with such violence that the mixture becomes incandescent. Well-dried carbonic acid is also without action on hydride of potassium, but in the presence of a slight quantity of steam it produces a formate. It is the same with many other bodies — ammonia gas and hydrochloric gas, for example, which ordinarily combine with the emission of thick white fumes, but do not do so after having been carefully dried. It will be remembered that by adding to dried salts of quinine traces of water vapor, they become phosphorescent and radioactive.*

On closely examining the role of bodies whose mere presence influences reactions we note that they behave as if energy were transported from the catalyzing body to that catalyzed. This fact can hardly be explained unless by acknowledging that the catalyzing body is undergoing the commencement of atomic dissociation. We know that, by reason of the enormous velocity possessed by particles of matter during dissociation, considerable quantitites of energy can be produced by the dissociation of a quantity of matter so small as to elude all attempts to measure it. The catalyzing substances could therefore be simply liberators of energy in matter on the earth and in all the universe. As in the atomic pile the fission material is worn out, so also platinum black and the colloid metals are eventually worn out — that is to say, by use they lose a great

***The Evolution of Matter** - *LeBon*

part of their catalyzing action. The theory involved indicates that all matter and energy simply represent a state of equilibrium between the internal elements of which the substance is formed and the external elements acting upon it. If this connection is not plainly apparent in some bodies, it is because they are so constituted that their equilibrium maintains itself without perceptible changes within the limits of fairly large variations of the medium. Water can remain liquid in variations of temperature ranging from zero degrees to 100 degrees Centigrade, and most metals do not appear to change their state within still wider limits of heat or of rates of vibration. However, these facts do not answer all the questions.

A succession of changes will be accompanied by the liberation of a certain quantity of the intra-atomic energy contained in matter. This is going on all the time in the cosmos to such an extent man with his man-made devices can never hope to compete. So why not use "nature's gift" of cosmic reactions? These actions, the mere presence of which are of such importance in the phenomena of life, may perhaps find an explanation in this theory. It was such studies on phosphoresence which led men to this hypothesis. It will be recollected that pure substances, various sulphides, phosphates of lime, etc., are never phosphorescent normally and only become so when brought to a red heat for a length of time with traces of various other substances — such as bismuth, manganese, etc. On the other hand, this elevation of temperature always provokes a dissociation of matter. It is therefore reasonable to suppose that the elements proceeding from this dissociation have an active part in the compounds then formed, which gives to such bodies the capacity of phosphorescence and sometimes other properties.

The combinations so obtained have precisely the characteristics pointed out above as belonging to extreme mobility; that is to say, of disintegrating and then regenerating themselves very rapidly. A ray of blue light falling on a screen of sulphide of zinc illuminates it in the tenth of a second, and a ray of red light falling on the same screen disinteregrates the phosphoresence in the same space of time; that is, it brings the screen back to its original state. These two contrary operations necessarily imply two converse reactions which may be indefinitely repeated.

These facts prove that by reason of the enormous quantity of intra-atomic energy contained in matter, a loss of substance too small to be detected by an analytical balance may be accompanied by a very great liberation of energy. We have no need to do this artifically because it is being done for us in the cosmos.

It is possible even without the action of heat to verify in ordinary bodies the existence of a constant emanation from the dissociated matter, though this emanation normally is extremely small in quantity. This all proves that the evolution of creation is going on continually.

To cause it to be apparent, it is necessary to compel it to accumulate in a restricted space. To demonstrate this, fold a sheet of metal so as to transform it into a small cylinder, similar to the one which encloses the ball of a condensing electroscope. With the lower opening closed, it is left for 8 days in darkness. Then, still keeping it in darkness so as to avoid any possible influence from light, the cylinder is placed on the insulating disc of the electroscope, or special ray counter instrument, to examine its radioactivity. After having charged the whole system it will be found that a definite discharge per minute is obtained. As the metal rapidly loses that which it has accumulated, the discharge soon ceases. Many materials other than metals, such as a box-wood cylinder, will produce the same effect; also certain gas-filled tubes.

The metal, after ceasing to act on the electroscope or counter, still has not exhausted its provision of radioactivity. It has simply parted with the quantity it can emit at the particular temperature at which the operation was effected. As with phosphorescent bodies or radioactive matter, it only has to be slightly heated to cause it to again yield a more considerable emission of active effuvia. The only difference between kinds of matter is in the rate of vibration or wave-length, the difference resulting in transmutation, changing and restoring. Evolution in its true sense is evolution from one rate of vibration to another, or transmutation: energy changing into matter and matter into energy, as we wrote back in 1925.

Cosmic rays are constantly creating radioactive carbon, and the fusion of small atoms to make larger ones gives off more energy than so-called "splitting" of the larger uranium or plutonium atoms, the uranium energy release ratio being only 1 to 1000 — that is only 1/1000 of the heavy atoms changed in the uranium "atom splitting" process.

"Nuclear fission" action is not confined to highly radioactive substances. "Atom splitting" is taking place naturally on this earth and on other planets and suns in the universe. There is absolutely no need of the hysteria that has swept the world over "atom splitting", it has been going on naturally from the very beginning of time. Energy and matter cannot be considered strangers just being introduced to the world. When the evolution of matter and the evolution of forces are considered, nothing new is being advanced. When we consider radium giving off three distinct forms of energy—alpha, beta, gamma—we cannot help but see a relationship between electricity and matter. Alpha ray? Alpha particle (matter). Beta ray? The electron (but still matter).

Radium decays into other nuclei (radon gas) which emit emanations that have characteristics of energy levels higher than those emitted by the original atom. This gas generating alpha rays could be condensed into a liquid if volume would vary inversely with the pressure (Boyle's Law). Thorium decays in much the same manner giving rise to higher energy products.

Now, if we turn to vegetable matter, we find energy is also being given off in a way similar to that of radium, thorium and other radioactive matter. This time, however, we find direct electrical energy being given off. Various vegetables and fruits will do for the experiment. A "volta pile" can be made of potatoes, grain, apples, onions, etc. Let us take the apple. In a "volta pile" made of 25 apples (50 halves) one can obtain enough electromotive force to light a flashlight globe. The living animal cell, like the vegetable cell, is also giving off energy, consuming oxygen and producing carbon dioxide and an electrical potential.

We find in vegetable, mineral and animal matter the same relationship, the same dependence on oxidation and electrical energy as we find in oxidation in stellar space, evolving energy into matter and matter into energy. Here we see energy and matter and matter and energy vibrating together.

Regarding the force of energy from the cosmos, noted experts in photography have found the light produced from this source of energy is much whiter than that obtained from ordinary electrical energy. This light burns into the film when photographed due to the great intensity of the pure white light produced, yet the light is easy on even weak eyes.

It is also noted that detail from these films can only be obtained by holding back dense areas to about one tenth that of normal exposure even when non-halation film and super flash bulbs are used, but the pictures are sharper. Let us again state: An electrical generator is in the true sense not a generator. It creates nothing. Electricity is not made by the generaor, it is merely captured or pumped. From that standpoint an electric generator is an electric pump, and a Radiant Energy or cosmic energy device, a high speed oscillating energy resonator.

If we reverse our imagination on what the telescope has taught us of the stellar universe, we will find that beyond the microscope we have the particles of which everything about us consists, obeying every law that is found everywhere from stellar space to atomic space. We find bodies in motion and when we think of how small the proton and electron are, and yet obeying the same laws of the universe, we may see the economic wastefulness of science trying to crack the atom, when nature, or call it what you will, is accomplishing the same thing for us in stellar and inter-stellar spaces. Why do something nature is already doing for us? Let's use what nature offers, as I first suggested many years ago. From this conception, we might see that Democritus came close to a great scientific truth when he boldly declared that all physical phenomena reduce themselves to one thing—motion—or as we might repeat, "vibration," "the sounds of the universe." Let me repeat: do not forget the atom is but a counterpart of the universe itself, and that light and other radiations exert a mechanical pressure upon every object they strike, and that all these radiations are essentially electrical in their character. There is a breaking down and a building up of the atom continuously, and it is this evolution which is producing, eternally, unlimited power.

May it not yet be shown that the dissipated energy which results from so much transformation of matter which has heretofore been unavailable — or should we say, unused by us where it has only appeared to be unavailable — should now become available to us, an unlimited source of power through the discovery of the application of these forces. And that the discovery that matter and energy are one is possibly the sum total of all that has been found during the centuries of constant research? Are the nature of matter and energy to be judged by that small portion of the universe which is visible to man — man still limited even when armed with his most powerful telescope or with his finest microscope?

All space is saturated with energies which are vibratory in their ultimate analysis and very closely allied to electrical action. The relation of matter to energy and energy to matter then becomes the potential of the universe, one continous series of oscillations, moving to and fro like a great pendulum across the universe. A steady flow of energy can be had from the surges of the universe, just as a steady flow of water may be obtained from the surges of the sea.

Electrons are spontaneously being emitted from all nuclei found in nature, and every new discovery on the subject bears out the claim that all "space" is filled with energy, millions of amperes at very high voltages.

This is based on the disintegration of radioactive material, natural or artificial; but the same action is taking place in the "sonics" (vibrations) of the universe with all matter and energy.

As long as the universe has existed, charged particles now called "cosmic rays" have been bombarding every planet or object in it, including all living bodies, at the rate of twenty times per second and with a force great enough to penetrate deep into the rocks of the earth.

Cosmic rays have been called "the death cry of the universe" or "by-products of the destruction of matter in the cosmos." In the process of the evolution of matter and evolution of forces there is no "death cry," only the "song of creation." There is no destruction of matter or energy, only the cycle of matter and energy. Creation is going on continually. You subtract in one place only to add in another.

CHAPTER 10

CAPTURE OF ENERGY BY RESONANCE

with the Radiations of the Universe

Oscillating Discharge - When any elastic substance is subjected to strain and then set free, one of two things happens. The substance may slowly recover from the strain and gradually attain its natural state, or the elastic recoil may carry it past its position of equilibrium, and cause it to execute a series of oscillations. Something of the same sort may also occur when an electrified condenser is discharged. In ordinary language there may be a continuous flow of electricity in one direction until the discharge is completed, or an oscillating discharge may occur; that is, the first flow may be succeeded by a back-rush, as if the first discharge has overrun itself and something like recoil has set in. The condenser thus becomes more or less charged again in the opposite sense, and a second discharge occurs, accompanied by a second back-rush, the oscillation going on until all the energy is either radiated or used up in heating the conductors. However, in the case of Radiant Energy, the oscillations can continue forever because of the actions of the universe. It is known that high frequency currents may be classified as to their oscillating characteristics, i.e., damped or undamped. The energy from the universe being of both types, depending on conditions beyond the scope of this writing, we have a back-rush effect as explained before in the RE device.

The purpose of the capacitors in the RE circuit is to store and then to discharge energy. If the resistance is low, an oscillatory discharge takes place. The discharge and recharge on the opposite plates continues in the ordinary sense until the energy which was originally stored is dissipated. In the case of the RE device, a different effect is established because the oscillations from the universe trapped by the RE valve, continue to enter the circuit in the way waves of the sea beat upon the shore. With negligible resistance in the device, no energy is lost in heat and the oscillations continue.

With the explanations given, is the obtaining of energy from the energy oscillatons of the universe any less understandable than the obtaining of energy from a mechanical prime mover? To use mathematical language, exact conditions exist in both cases, be they mechanical or electrical systems. The electrical prime mover, the so-called electric generator or the Moray energy oscillator all have the same scientific facts. In a mechanical system, the greater the inertia the greater the tendency of the body to keep in motion once it is set in motion. In an electrical circuit, the greater the inductance with resistance cut to a minimum, the greater the tendency of the electrical energy (current) to continue to flow once it is started.

From the above it then becomes clear that if

$$R < \sqrt{\frac{4L}{C}}$$

where R is the resistance in ohms, L is the inductance in henries, and C the capacity in farads, oscillating discharges take place. For low resistance value, i.e., R, the frequency of the oscillations may be expressed.

$$f = \frac{1}{2\pi\sqrt{CL}}$$

Let Q be the charge of the capacitor at any instant, C its capacity, R the resistance of the circuit, and L its coefficient of self-induction. Then, if I is the intensity of the current and E the electromotive force, we have the equation:

$$E - IR = \frac{d}{dt}(LI) = L\frac{dI}{dt}$$

In this case E = Q/C, and I = dQ/dt therefore:

$$L\frac{d^2Q}{dt^2} + R\frac{dQ}{dt} + \frac{Q}{C} = 0.$$

The solution of this equation is:

$$Q = Ae^{\mu t} + Be^{\mu' t}$$

where μ and μ' are the roots of the equation.

$$\mu^2 + \frac{R}{L}\mu + \frac{I}{CL} = 0$$

or

$$\mu = -\frac{R}{2L} \pm \sqrt{\frac{R}{4L} - \frac{1}{CL}}$$

Writing

$$\alpha = \sqrt{\frac{R^2}{4L^2} - \frac{1}{CL}},$$

we have

$$\mu = R/2L + \alpha, \text{ and } \mu' = -R/2L - \alpha,$$

and

$$Q = e^{-\frac{Rt}{2L}}(Ae^{\alpha t} + Be^{-\alpha t})$$

where A and B are constants determined by the initial conditions; viz. that initially we have Q = Qo, and I = 0, which give

$$A + B = Q_o, \text{ and } A\mu + B\mu' = 0,$$

or

$$A = Q_o(\tfrac{1}{2} + \frac{R}{4L\alpha}), \text{ and } B = Q_o(\tfrac{1}{2} - \frac{R}{4L\alpha}).$$

Hence, at any time we have

$$Q = Q_{oe}^{-\frac{Rt}{2L}}\{(\tfrac{1}{2} + \frac{R}{4L\alpha})\, e^{\alpha t} + (\tfrac{1}{2} - \frac{R}{4L\alpha})\, e^{-\alpha t}\}.$$

Consequently the current at any instant is

$$I = -\frac{dQ}{dt} = \frac{Q_o}{2CL\alpha}\, e^{-\frac{Rt}{2L}}(e^{\alpha t} - e^{-\alpha t})$$

Hence if α be real — that is, if we have

$$R^2 > 4L/C$$

the quantity Q will gradually diminish to zero as the time increases. If, however, we have

$$R^2 < 4L/C$$

then α will be imaginary, and writing

$$\alpha' = \alpha\sqrt{-1} = \sqrt{\frac{1}{CL} - \frac{R^2}{4L^2}}$$

the above formula becomes at once

$$Q = Q_{oe}^{\frac{-Rt}{2L}} (\cos \alpha' t + \frac{R}{2L} \sin \alpha' t)$$

and

$$I = \frac{Q_o}{CL\alpha'} e^{-\frac{Rt}{2L}} \sin \alpha' t.$$

In this case the current starts from zero and rises to a maximum; it then falls to zero and becomes reversed, after which it passes through a series of oscillations. The discharge therefore does not take place in a single flow from one capacitor to the other, but a back-rush sets in, and a series of currents, or oscillations, occur alternately in opposite directions.

The current attains its maximum intensity when

$$\tan \alpha' t = 2L\alpha'/R \quad \text{(maximum current)}$$

The zero value of the current is reached when

$$\alpha' t = n\pi \quad \text{(Zero Current)}$$

and consequently the charge at the same time is at its maximum, for we have $I = dQ/dt$. Thus the charge oscillates backwards and forwards, attaining positive and negative maxima after the lapse of equal intervals π/α', the time of a complete being

$$I = \frac{2\pi}{\sqrt{\frac{1}{CL} - \frac{R^2}{4L^2}}}$$

If the resistance is small compared with the reciprocal of the capacity we may use the approximate formula

$$T = 2\pi\sqrt{CL}.$$

The successive maximum charges occur when $I = O$, or $\alpha' t = n\pi$; they are therefore

$$Q_o, Q_1 = -Q_{oe}^{\frac{-\pi R}{2L\alpha'}}, \; Q_2 = Q_{oe}^{\frac{-2\pi R}{2L\alpha'}}, \; Q_3 = -Q_{oe}^{\frac{-3\pi R}{2L\alpha'}}$$

The quantities therefore diminish in geometrical progression, and the energy of the charge diminishes correspondingly on each oscillation, being lost in heating the circuit.

Whether the discharge is continuous or oscillatory therefore depends on whether 4L is less or greater than CR^2, and an oscillatory discharge may be obtained either by increasing L or sufficiently diminishing C and R.

These predictions on analysis have been confirmed, as Thomson suggested, by examining the spark, during discharge, by means of a revolvig mirror. In Feddersen's experients the image of the spark in a revolving mirror was viewed through a telescope. When the resistance of the circuit was high, the spark was merely drawn out in width — that is, at right angles to its length; but when the resistance was sufficiently reduced, so that the oscillating discharge might occur, the band was reduced to a broken image consisting of a series of strips, each strip corresponding to a discharge. As stated, we have discovered the oscillation of the universe corresponds to oscillation of electric capacitors, depending on the frequency of the energy involved, going on into infinity. The arithmetic progression becomes geometrical.*

These oscillations of energy surging from the universe into the device and returning to the universe are picked up by the Moray device and the oscillator tubes of the device because it is tuned to oscillate in harmony with the oscillations of the universe — just as vibrations from musical instruments can be made to oscillate together. Every oscillation, whether large or small, is completed during the same interval of time. The beat note of time, the heartbeats of life, the oscillations are all governed by the same cycle of time; or as stated above, completed during the same interval of time; and as stated years ago, these waves of energy have a regular beat note of time, coming and going as the waves of the sea, but in a very definite mathematical order of time. These waves of energy come to the earth from every direction, stronger in the daytime than at night, but always coming with regular beat notes that might be referred to as the Father of Time, the Sire of Gravitation.

That matter and energy are possibly one is the sum total of all that has been found during three centuries of incessant research covering that portion of the universe visible in a forty inch telescope armed with the most powerful spectroscope ever made.

It is the experience of the writer that all space is saturated with inconceivably minute corpuscles of energy. Reference is made to the discoveries of Professor J.J. Thomson. These are doubtless either electricity in its ultimate refinement or very close to it. The earth and sun, all suns and dark bodies in space, all granular matter move through the primordial cosmic mass of electrical energy as a wire screen through water. The wide spaces in diamond, glass, steel, flint or anything else allow "these bodies smaller than atoms," as Thomson says, "to pass through."

From the definition of energy, it is the potential of the universe. When matter is in a phase allowing it to be active, it affects other quantities of matter at a distance. The method of transfer is known to be by means of wave and corpuscular motion. Each impulse moves from the omitting to the receiving mass on a rigorously straight line. One continuous set of oscillations in this line is called a ray. Each negative or "Thomsonian corpuscle" makes a double vibration to and fro like a pendulum straight across the direction of the ray — i.e., at right angles to it, the corpuscles moving and returning to the original position.

**Appendix II removed at this point*

Since the corpuscles are negative and can be drawn out of their original straight path by the action of magnetism, the entire wave motion of the universe is electromagnetic. This is what Maxwell theorized many years ago. Thomson proved the theory a fact. He wrote:

"After one corpuscle makes an oscillation across the direction of the ray and returns, the next does likewise, and the next, and so on. After the first corpuscle makes the swing, another distant from it 186,000 miles in the same straight line will also make a vibration at the end of the first second of time."

Quoting from **Atoms and Rays,** written in 1924 by Sir Oliver Lodge, Fellow of the Royal Society of Science and holder of five or six doctorates, with honors from scores of colleges and president of a dozen or more scientific societies:

"The term 'light', strictly speaking, means that kind of etherial radiation which is able to affect the eye. But it is common knowledge that there are many other varieties of radiation besides these to which the eye is sensitive. It is not clearly known why the eye is sensitive to some kinds of etherial radiation, and not to other kinds. That, no doubt, can be ascertained; it is a question for physicists and physiologists in collaboration. But the eyes of animals and insects as well as of man, all appear to be sensitive to a limited range of etherial radiation, which is therefore called light. Other kinds of radiation can affect a photographic plate; other kinds, again, can stimulate the chemical actions going on in the leaves of plants, and thereby supply the energy needed for vegetable growth. Another kind — a rather deeper harmonic as it were — supplies everything on earth with warmth, and by evaporating water contributes to most of the phenomena of weather. Other kinds, again, are emitted when individual electrons traveling at a high speed in a vacuum, encounter the obstruction of a target; this kind of invisible radiation being called x-rays. And, at the opposite end of the scale, another kind of radiation is emitted and is known as the Hertzian waves employed in radio.

"In speaking of these kinds of radiation as different we are not speaking quite accurately. They differ only as treble notes differ from bass notes; they differ in rapidity or rate of vibration, wave-length, or frequency. They do not differ in any other essential particular. Through the whole range — from the telegraphic waves, which may be a mile long, to x-rays, of which the wave-length is actually smaller than atoms, and only expressible in billionths of an inch — they all travel, we assume, at precisely the same speed. They are all of the same electgromagnetic character; they are all subject to the same laws of interference, of reflection, refraction, and polarization, which have long been studied in the department of physics known as optics."

(T. Henry Moray would qualify the above statement by saying there are speeds greater than 186,000 miles per second.)

"Whatever an electric charge is, or is not, it is certainly a focus of energy. And if we could imagine a vortex, containing the known mass of the electron and circulating with the velocity of light, its energy would be equal to that of the electric field in the space surrounding the electron. This coincidence, if it be a coincidence, can hardly fail to have some meaning. And there are those who are beginning to think that the whole material universe is built up of energy in various states of self contained or intrinsic motion; by which adjectives it is intended to discriminate between rotary motion (like that of a top or a whirlpool) and ordinary locomotion (or shifting from place to place.)

"Those who hold this view of the universe are strengthened in their position by the statements of Einstein on energy in general. It is well known that all the ordinary energy we are acquainted with, such as the motion of railway trains, cricket balls, and such like, is merely relative — relative to the earth, or to some other piece of matter. There is nothing absolute about it. But Einstein gives an expression for what one might be inclined to call absolute energy, in which the only relevant velocity is the velocity of the Cosmos. And all the phenomena we observe in nature, at any rate in inorganic nature — omitting the phenomena of Life and Mind for the present, as lying outside our physical ken — may be regarded as due to, and as demonstrating, slight modifications of the portion affected by this great velocity, in form which enables it to appeal to our animal-derived senses. For the spinning motion itself is impalpable and beyond the ken of our instruments, until it partially exhibits itself as transmitted waves in the form of radiation.

"All the energy that we experience can be resolved into vibrations or tremors. But all electric and magnetic phenomena, and therefore, all chemical activity, are likewise known to be modes of manifestaion of vibrating space, the complete manner and meaning of which have still to be worked out.

"So the question arises, What is Matter? Is that too a manifestation of some peculiar properties in the medium? We know now that matter is built up of protons, neutrons, electrons, etc. But when we come to analyze these into their fundamentals, we find far more than a hint that they are but special modifications in the all-pervading energy, and are essentially resolvable into kinetic energy of a specific kind. Hence, we are beginning to think that all matter itself is a form of energy.

"Energy is the chief thing in the physical universe that directly appeals to us. We apprehend it under a great variety of forms. And it is becoming probable that what we call matter is one of these forms. Most of the forms of energy that we know are convertible one into another. The energy of motion turns into heat; so does the energy of electric currents, unless it is converted into the energy of chemical separation or electric charge. Conversion from one form to another, without loss, is the sign-manual of energy. And the proof that matter is a form of energy will not be clinched until it can be demonstrated that matter too is convertible into other forms of energy."

Many attempts have been made to harness the forces of nature directly to produce usable power.

Other types of energy besides sunlight are constantly bombarding the earth with waves and particles, or surrounding it in much the same manner as its atmosphere or the very space in which it exists. Among these are the gravitational and magnetic fields and the cosmic rays.

The features of cosmic radiation would enable RE transformers using cosmic radiations as a primary source to operate with relative independence as to position or season. It is also evident that such a device would show possibilities of efficient operation in moving vehicles within the earth's atmosphere and in space at continuous high power levels.

Such an energy transformer or converter has been built. It has been operated, at full load continuously with no expenditure of fuels of any type, without a mechanical prime mover, kept alive by the oscillations of the energies from the cosmos; an energy converter, or transformer, which would be capable of converting the high frequency, high level energy of the cosmic radiation into current of usable frequency and voltage.

Basically the theory of operation is as follows: Oscillations are started in the first stage or circuit of the device by exciting it with an external energy source. The circuit is "tuned" until the oscillations are sustained by harmonic coupling to the cosmic wave frequencies. The reinforcing action of the harmonic coupling increases the amplitude of the oscillations until the peak pulses "spill" over into the next stage through a special detector or valve which prevents the return or feedback of energy from succeeding circuits. These "pulses" drive this stage, which oscillates at a lower frequency and is again reinforced by harmonic coupling with the ever present cosmic waves. The second stage drives a third stage, and additional stages are coupled until a suitable power level at a usable frequency and voltage is obtained by means of special transformers. (*See schematic drawing of RE device.)

Once the machine is in operation and delivering power, it does not require any continuance of the excitation forces necessary to start it. The oscillations are sustained as long as it remains properly tuned and the external circuit is completed through a suitable load.

The special tubes which appear to be the key to the success of this device are ionic cold cathode tubes which require no external power sources.

Practically speaking, the present method of securing energy with prime movers has been used too long. We have used it so long that we have let ourselves believe that there is no other way to secure energy. Present methods are expensive and cumbersome. Think of the power sites and steam plants, the transmission lines, the transformers, the many other kinds of equipment that are now required to supply electricity. The present system is unreasonably expensive. When we begin to realize the tremendous energies, electrical energies, that surround us, we should actually be ashamed that we have been so dilatory in our efforts to improve upon our crude, cumbersome, and expensive method of furnishing heat, light and power.

Books could be filled with accounts of the struggles for acceptance of a new idea or a new way of doing things. You need only read the story of the development of the telephone, the railroad, the airplane, automobile, to appreciate the cost in sacrifice, money and unselfishness necessary to promote a new and revolutionary movement, irrespective of the vast good it accomplishes when established for the use of humanity.

Schematic Drawing

Diagram: Flow-Sheet Showing Different Stages
The RE Goes Through

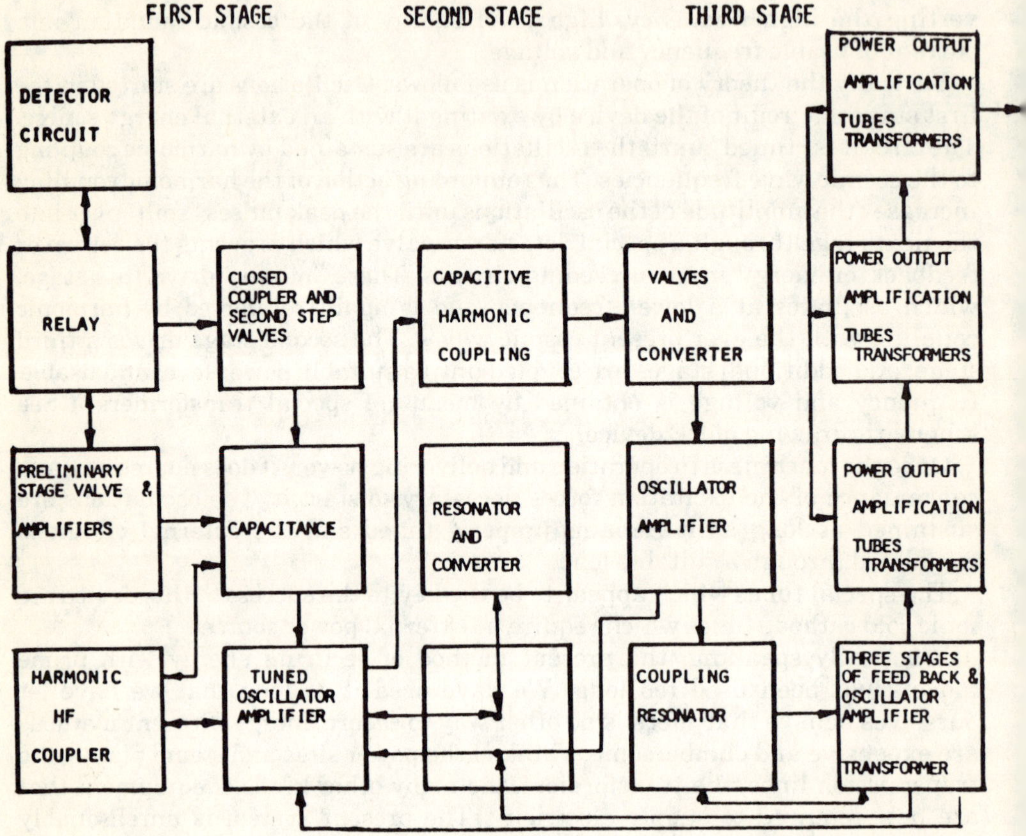

CHAPTER 11
CONCLUSION

We have attempted to give a complete history of the Radiant Energy device as developed by Dr. Moray and described in his book, **The Sea of Energy,** first printed in 1931. We have corrected typographical errors, removed repetitious material, and made minor editorial changes in the text of the original edition. Henry Moray, of necessity, generated his own terminology for his original concepts, a terminology that is not always the same as that in general use today. Most of his theory, developed in the 1920s and '30s were innovative at the time. For example, one engineer pointed out to me recently that when Henry Moray spoke of "cosmic rays" and of "energy from the cosmos," he was not necessarily using the terms synonymously.

Some of Moray's contemporaries claimed accomplishments similar to those we have discussed here. Readers may be familiar with the claims that Alfred Hubbard of Seattle demonstrated a device in 1919 that transmitted power without wires. Lester Jennings Hendershot claimed to have a revolutionary free-energy motor in 1925-28. In 1927 a Denver newspaper reported that Fred Walden and E.G. Lewis had demonstrated 200 watts of energy drawn "from the air." In Michigan three years later, Chancey J. Britain was said to have produced an undisclosed wattage from his device; enough to power his own home.

As far back as 1901 Nicola Tesla is said to have patended free energy devices, U. S. patent numbers 685957 and 685958, described in the book **Nicola Tesla: Lectures, Patents, Articles,** published by the Nicola Tesla Museum, Belgrade, Yogoslavia, in 1956. As a member of the Tesla Society, my father obtained two copies of the book, but he had no access to or knowledge of the patents before the book was published. They could in no way have influenced his own research on RE.

Research into sources of free energy are continuing today. I list here the names of those I know have worked in the free energy field. Others I would appreciate hearing about, or if the experimenters are living, I would like to hear from them.

— Edwin B. Gray of Los Angeles has received publicity lately. I have not been able to contact Mr. Gray directly, so I only know what has been written in various magazines.

- F.R. Reeton from Yorkshire, England (151 Cold Bath Road, Harrogate), believes that he can build a "resonant cord inductor" electrical machine.

- Bob Jones and his entropy engine is mentioned in a booklet printed by James E. Cox, 1972, P.O. Box 793, Pamona, CA.

- Bruce E. Depalna of Keller Church Road, Bedminister, PA., and his partner, Ken Sniderman, have developed a device they claim gives more energy than it uses. I met Depalna recently in Philadelphia.

- Bob Teal, box 802, Madison, Fla., reports that he has a motor that operates at very high frequency. The secret is that he has connected this device through an oscillatory circuit and completed the circuit from air to ground. I asked him why he did not use a solid state device and bypass the moving parts. He said he was able to do this; however, he was unable to get more than a few watts of power.

- Wilhelm Reich, now deceased, developed a motor that I understand he claimed to use "orgone" energy. This same orgone energy was supposedly used as a therapeutic treatment, which I believe is still in use today.

- Leonard Bergstrand, Rt. 2, Vinton, Iowa, has developed a motor using magnetic crystals of some kind. This device was rumored to have been purchased by Allis-Chalmers Corporations with the stipulation that Bergstrand discontinue research in the field.

In all cases it would seem that anyone who develops such a device also develops considerable opposition. The device is eventually either purchased by some large corporation for unexlained reasons and never brought to market, or the device is lost, or, as in the case of Edwin Gray, the inventor becomes tied up over secondary squabbles with security exchange commissions or other bureaucracies that have nothing to do with the actual operation of the device itself, but only with the business arrangements that have been made to produce it.

As far as I know, fraud charges are seldom leveled, and although I cannot vouch for any of the devices I mention above, I can't believe that they all can be completely worthless, having seen my father, Henry Moray, work on his own device.

The question with any of the devices is this: How much power can it produce under controlled experimental conditions? The following guidelines delineate what one must do to duplicate Dr. Moray's Radiant Energy Device:

1. When the primary side of the device is disconnected, a spark equivalent to 225,000 volts must be drawn between the two disconnected terminals.
2. By disconnecting the antenna and re-connecting immediately, the lights must stay on. However, if the lights are allowed to die, re-connecting the antenna will not re-establish the connection and the unit becomes completely electrically dead.
3. The energy produced must be high frequency.
4. The device must be operable at a distance of more than 50 miles from power lines or radio stations.
5. The light of a 100 watt lamp must be whiter and brighter than the ordinary 100 watt lamp without burning out after 157 hours of operation.
6. When a heavy load is connected to the device while in operation, after already draining as much as 4 kilowatts, the present load must not decrease nor the lights flicker.
7. A duplicate shunt test, like that made by Lovesy, with number 50 wire must be successful.
8. Small number 30 wire should be used for all circuits.
9. At least, 10,000 watts, and up to 50,000 watts of power, must be produced

in a box weighing not more than 60 pounds.

10. As the ground wire is driven deeper, the amount of power must increase.
11. None of the components of the device must heat.
12. No moving parts may be included.
13. The device must be absolutely silent.
14. By tuning the circuit by the use of variable capacitors, the lights must come on.
15. A resistive load of 600 watts must be brought to full heat in less than 2½ minutes.
16. After 158 hours of operation, the unit must still be running and no parts of the device be heated.
17. The test performed by Dr. Knudson must be duplicated.
18. Smell of ozone should be present during the operation.
19. No heated cathodes or bias power supply must appear.
20. All of the above must be present at the same time.

The fact remains that Henry Moray demonstrated under the above conditions on numerous occasions, up to 50,000 watts of power. The device always sat on a table where it was easily examined by anyone coming into the room, so that it could be seen that the only wires entering the device were the antenna and the ground.

Today, attempts are made to give the impression that this is not correct. Descriptions given by witnesses along with photographs, more than adequately establish that this was the case. My brother and I are of the opinion that for whatever reasons, whether personal or business, at one time or another many of te individuals who became associated with Radiant Energy have attempted to discredit Henry Moray without regard to what the results may have been. We are convinced that Dr. Moray's fears that he had been betrayed by those whom he took into his confidence, may not be groundless.

The Cosray Research Institute has but one purpose in mind and that is to insure that Dr. Moray's work is continued. In an attempt to raise funds, we have formed the House of Moray Research Foundation, which we hope will be the means to develop Radiant Energy and other discoveries that promise to be of great importance in the progress of mankind. We hope that the reader will aid in the work by encouraging others to read this book.

We wish to invite those who have either discovered an energy source similar to Radiant Energy or any other energy source that is unexplained to unite with us. Only by working together are we going to be able to overcome the opposition to this field of research.

History has adequately demonstrated and established that for one to build for man's benefit a working model of a concept revolutionary to science will not necessarily enduce the world to clamor for it (regardless of how great or maybe even because of how great its potential is). Quite to the contrary, it only opens the door to extremes of skepticism, ridicule or even persecution from his peers.

Not until such a new item has been standardized and provided in numbers to many (quite often during severe opposition and at overwhelming expense), do the brave men adopt and enlarge its utility thereby lessen the risk of its being totally suppressed.

We believe the only way mankind will be given the opportunity to benefit from Radiant Energy is by the widespread dissemination of the notes combined with a collective effort by many interested parties to standardize it. In this way safety in numbers is achieved. For this reason we propose the following plan and encourage financial and supporting assistance to bring Radiant Energy to practicality for the benefit of mankind.

Cosray Research Institute has recently compiled all of Dr. Moray's notes. If funding were available, we believe these notes would allow the reproduction of his work. The question is, what can be done to insure the most expeditious means of developing and disseminating this discovery for the benefit of all mankind? We propose to go about it in two ways.

First, the House of Moray Research Foundation has agreed to offer $1,000 for the best theoretical paper explaining Dr. Moray's work. Many people seem to think that they have to explain theoretically what is happening before they can accept the evidence that it does happen and is reproducible. The fact that none of the researchers in the field have had exact details of how Dr. Moray did his work, and consequently have not reproduced it themselves, is of no importance to us. However, it does seem to be of some great importance to the rest of the world to see how it could have worked, and therefore, in order to develop this field of research, we have concluded that it may be an advantage to approach this research first from the theoretical standpoint.

Contestants for the best theoretical paper explaining Dr. Moray's work must obtain an application from COSRAY, 2505 South 4th East, Salt Lake City, Utah, 84115. The theoretical approach must deal with the phenomena as described by Dr. Moray, although other phenomena not directly connected may be considered.

Second, and from the more practical standpoint, the Cosray Research Institute hopes to interest persons to join with it in the capacity of consultants, with full use of all information now available on Radiant Energy. Cosray is hereby offering a cash prize of at least $10,000 and up to $100,000 (depending on the number of participants) to the consultant who submits to Cosray Research Institute a report and working model which in the opinion of Cosray makes the greatest contribution in advancing the standardization of Dr. Moray's Radiant Energy Device. The decision as to which presentation and model submitted, if any at all, qualifies for the prize, rests solely with Cosray Research Institute.

In order to maintain proprietory rights, Cosray Research Institute will enter into a consultation contract and will lease to the consultant for eighteen months, copies of Dr. Moray's notes, with comments as to the result of research made by Cosray in attempting to reproduce the device. We will include additional notes made by Cosray in its own recent investigations in which Cosray spent nearly $450,000 to collect and correlate Dr. Moray's notes.

To participate in the consultation project, write to COSRAY at the above address. We will send an application with a self-addressed, stamped envelope; return this application to us within 30 days. Cosray Research Institute will review the applications and, at its discretion, send an unexecuted consulting

NAME:

ADDRESS:

PHONE NO.:

Place
Stamp
Here

COSRAY RESEARCH INSTITUTE
2505 South 4th East
Salt Lake City, Utah 84115

NAME:

ADDRESS:

PHONE NO.:

Place
Stamp
Here

COSRAY RESEARCH INSTITUTE
2505 South 4th East
Salt Lake City, Utah 84115

Gentlemen:

I wish to participate in your contest on the theoretical investigation of how Dr. T. Henry Moray's Radiant Energy device may have worked. I understand that my paper must be submitted prior to January 1, 1980.

Signed,

NAME ______________________________
(Please Print)

ADDRESS ______________________________
Number and Street

____
City, State, and Zip Code

This offer is void where prohibited by law.

Gentlemen:

I am interested in working as a consultant for Cosray Research Institute. Please forward an application to me as soon as possible.

Signed,

NAME ______________________________
(Please Print)

ADDRESS ______________________________
Number and Street

____
City, State, and Zip Code

This offer is void where prohibited by law.

agreement with the $1.00 consultation fee, lease agreement and escrow agreement to the applicant, who will then execute these documents and return them with the lease payment of $1,000 to Cosray — Randy B. Hart, Trust Account, at the above address. The lease payments will be held in an escrow account until at least $100,000 has been received, at which time the funds will be released to Cosray Research Institute and each consultant will receive a registered copy of all Dr. Moray's notes. The consultants will then have full use of the notes except as limited in the consulting and lease agreements. Cosray will lease at additional cost any components which it deems proprietory in nature.

Cosray Research Institute will have first option to utilize any information, presentation, or model submitted by any consultant within one (1) year from the date of submission. If Cosray fails to exercise its option within the prescribed period, title and all rights to the information, presentation, or model will remain with the consultant. Any consultant submitting any information, presentation, or model subsequently used to further develop RE, will receive a royalty, to be negotiated with Cosray Research Institute. Cosray may reject any presentation, information, or device submitted, thereby releasing it from any obligation to pay any royalty to the consultant.

Should Cosray Research Institute fail to establish at least $100,000 in lease payments from proposed consultants, all monies will be refunded less 7% for management, escrow and administrative costs.

This offer is void where prohibited by law.

REFERENCES

1. Pollard and Davidson, **Applied Nuclear Physics,** page 70
2. Dr. Robert Baker, **Astronomy,** page 303
3. Dr. Gustave LeBon, **Evolution of Matter**
4. Dr. Gustave LeBon, **Evolution of Forces**
5. Dr. Nikola Tesla, **Experiments with A.C. Currents of High Potential and High Frequency Experiments**
6. Dr. Edgar L. Larkin, **Radiant Energy,** page 17
7. "Element formation in the Stars is also energy transformation in the Stars," *Sky and Telescope,* July 1956, August 1956, March 1959, pages 241-243
8. *Scientific American,* May 1958, page 44
9. *Scientific American,* February 1960, page 53
10. *Scientific American,* June 1960, page 64
11. Dr. M. Luckiesh, **Foundations of the Universe,** pages 5, 14, 43
12. **Progress in Elementary Particle and Cosmic Ray Physics,** Vol. 7, North-Holland Publishing Co., Amsterdam
13. J.A. Eldridge, **The Physical Basis of Things,** chapters 3-6, McGraw-Hill Book Co., New York, 1934.
14. G.F. Hull, **An Elementary Survey of Modern Physics,** Macmillan, New York, 1936
15. L.B. Loeb, **Kinetic Theory of Gases,** McGraw-Hill Book Co., New York 1934
16. G.E.M. Jauncey, **Modern Physics,** D. Van Nostrand Co., New York
17. J. Perrin, Taylor & Francis, **Brownian Movements and Molecular Reality,** 1910
18. J.A. Crowther, **Ions, Electrons and Ionising Radiations,** Longmans, Gree & Co., New York, 1924
19. C.G. Darwin, **The New Conception of Matter,** Macmillan, New York, 1931
20. J.A. Eldridge, **The Physical Basis of Things,** chapters 9 & 10, McGraw-Hill Co., New York, 1934
21. W. Gerlach & F.J. Fuchs, **Matter, Electricity and Energy,** D. Van Nostrand Co., New York, 1928
22. A.E. Haas, **The World of Atoms,** D. Van Nostrand Co., New York, 1928

22. A.E. Haas, **The World of Atoms,** D. Van Nostrand Co., New York, 1928

23. G.E.M. Jauncey, **An Elementary Survey of Modern Physics,** D. Van Nostrand Co., Inc., New York, 1932
24. O.J. Lodge, **Atoms and Rays,** Doran, Inc., New York, 1929
25. R.A. Millikan, **The Electron,** Ginn & Co., Boston, 1924
26. B.A. Russell, **The A B C of Atoms,** E.P. Dutton Co., New York, 1921

27. S.G. Starling, **Electricity & Magnetism,** Longmans, Green & Co., New York, 1929
28. A.E. Haas, **Atomic Theory,** D. Van Nostrand & Co., New York
29. H.A. Kramers & H. Holst, translated by R.B. and R.T. Lindsay, **The Atom and the Bohr Theory of Its Structure,** Alfred A. Knopf, New York, 1924
30. A. Sommerfeld, translated by Brose, from third German edition, **Atomic Structure and Spectral Lines,** **Methuen** & Co., Ltd, London, 1923
31. J.a. Eldridge, **The Physical Basis of Things,** chapters 14-16, McGraw-Hill Book Co., New York, 1934
32. G.E.M. Jauncey, **Modern Physics,** D. Van Nostrand Co., Inc., New York, 1932
33. **The Advance of Science,** Watson Davis, 1934
34. *Scientific American,* 1930, pages 230-231
35. Physics Staff at the University of Pittsburgh, **Atomic Physics,** 2nd edition, John Wiley & Sons
36. J.J. Thompson, **Electricity and Matter,** 1904
37. Janet, **Lecons d' Electricite,** 2nd edition, pages 2 & 5
38. M. Ruderfer, "Neutrino Structure of Ether," **Lettere II Nuovo Cimento** Vol. 13, N. 1 & 9, 1975
39. H.C. Dudley, **The Morality of Nuclear Planning,** Kronos Press, 1976
40. G. Gamow, **Thirty Years that Shook Physics,** Doubleday, N.Y., 1966
41. J.A. Wheeler, et.al., **Gravitation,** W.H. Freeman and Co., 1970
42. J.A. Wheeler, **Geometrodynamics,** Academic Press Inc., 1962
43. T.H. Boyer, "Random Electrodynamics: The Theory of Classical Electromagnetic Zero-point Radiation", *Phys. Rev.,* D11, No. 4, 790, 1975
44. M.O. Scully, M. Sargent, "The Concept of the Photon", *Physics Today,* 38, March 1972
45. E.G. Harris, **A Pedestrian Approach to Quantum Field Theory,** Wiley Interscience, chapter 10, 1972
46. S.L. Adler, "Some Simple Vacuum, Polarization Phenomenology . . . ", *Phys. Rev.,* D10, No. 11, 1974
47. Procaccia and J. Ross, *Science,* 198, 716, Nov. 18, 1977
48. P. Glansdorff & I. Prigogine, **Thermodynamic Theory of structure, Stability, and Fluctuations,** Wiley Interscience, N.Y., 1971
49. *International Journal of Fusion Energy, Fusion Energy Foundation,* Vol. 1, No. 1, 1977; No. 3-4, 1978
50. W.H. Bostick, "Experimental Study of Plasmoids", *Phys. Rev.,* 106, No. 3, 404, 1957
51. D.R. Wells, "Dynamic Stability of Closed Plasma Configurations", *J. Plasma Phys.,* Vol. 4, 654, 1970
52. J.P. Walters, *Science,* 198, No. 4319, 787, Nov. 1977
53. P.O. Johnson, "Ball Lightning and Self Containing Electromagnetic Fields", *Am. J. Phys.,* 33, 119, 1964
54. Moray B. King, "Energy Source Implications of a Helicon Toroid Model for Ball Lightning", *QPR,* No. 18, Valley Forge Res. Center, Moore School, University of Pennsylvania, 1976

55. P.A. Silberg, "Ball Lightning and Plasmoids", *J. Geophys. Res.,* 67, No. 12, 4941, 1962
56. S. Singer, **The Nature of Ball Lightning,** Plenum Press, N.Y., 1971
57. A.C. Scott, et. al., "The Soliton: A New Concept in Applied Science", *Proc. IEEE,* Vol. 61, No. 10, 1443, Oct. 1973
58. S. Bardwell, "The Implications of Non-Linearity", *Fusion Energy Foundation Newsletter,* reprint
59. G.L. Lamb, "Solitons and the Motion of Helical Curves", *Phys. Rev.,* D14, No. , 235, 1976
60. F. Lund & T. Regge, "Unified Approach for Strings and Vortices with Soliton Solutions", *Phys. Rev.,* D14, No. 6, 1524, 1976
61. R.S. Knox, **Theory of Excitons, Solid State Physics,** Academic Press, N.Y. 1963
62. D.J. Bohm & B.J. Hiley,
On the Intuitive Understanding of Nonlocality as Implied by Quantum Theory", *Found. Phys.,* Vol. 5, No. 1, 93, 1975
63. H.P. Stapp, "Are Superluminal Connections Necessary?" *Il Nuovo Cimento,* Vol. 40B, No. 1, 1977
64. H. Evertt, **The Many Worlds Interpretation of Quantum Mechanics,** Princeton University Press, 1973
65. H. Evertt, "Relative State Formulation of Quantum Mechanics", *Rev. Mod. Phys.,* 29, No. 3, 454, 1957
66. B. Toben, **Space-Time and Beyond,** E.P. Dutton and Co., Inc., N.Y., 1975
67. "Tapping the Earth's Electric Field", *Machine Design,* April 15, 1971
68. "Background Cosmic Count Increase Associated with Thunderstorms", *Journal of Geophysical Research,* reprint, Vol. 72, No. 18, September 15, 1967

LIST OF APPENDICES

APPENDIX I
Deletions from original chapter 1, now chapter 7

Salt Lake Tribune, June 14, 1960.. Preliminary analysis of Explorer VI data just published shows the (radiation) belts shrunken in size and in intensity — at least this was the state of affairs during August and September of last year.

Another surprise comes in the discovery of a third belt lying between the outer and inner belts found by Van Allen. Perhaps this new belt should be called the Arnoldy, Hoffman, and Winckler belt in honor of the three University of Minnesota cosmic ray scientists responsible for its discovery.

The three physicists from the Land of Lakes masterminded the construction of the radiation detection equipment which formed part of the payload of Explorer VI and analyzed the data it accumulated.

The new shape in radiation belts found by the Minnesotans confirmed a growing suspicion that the belts are not the stable pair of doughnuts once visualized, but a shimmering set of tenuous rings which may change markedly from day to day.

Radiation levels measured by Pioneer III, Pioneer IV, the Russian Mechta, and Explorer VI have shown significant variations.

Fortuitously, Explorer VI was in orbit and busily taking radiation measurements during a strong magnetic storm created by unusual activity on the face of the sun.

As a result, the step-by-step reaction of the belts to such a storm and a rather complete history of their return to normalcy was obtained for the first time.

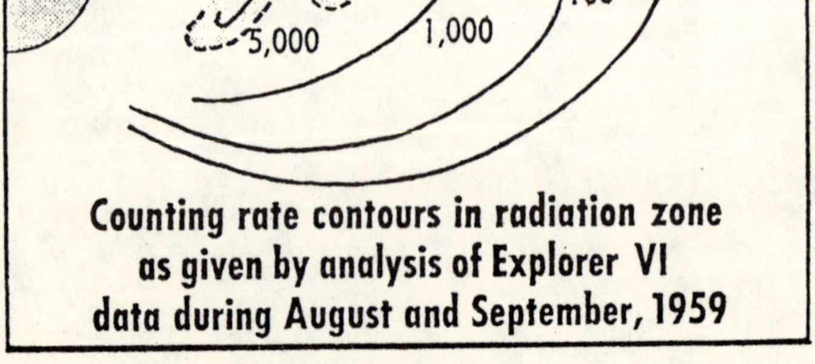

Counting rate contours in radiation zone as given by analysis of Explorer VI data during August and September, 1959

The orbit of Explorer VI was ideal in several respects for the probing of the intensity and extent of the Van Allen belts.

Its orbit was highly elliptical, extending from about 4,100 miles from the earth's center at its near point to some 30,000 miles at its far point.

Moreover, the inclination of the orbit was such that the rocket passed from rather high northern magnetic latitudes to high southern magnetic latitudes.

Because of these two orbital characteristics, a large portion of the region of the belts was scanned in several days time.

That portion of the new data which has been processed to date indicates that the maximum normal radiation dosages in the belts during the period of Aug. 7 to Oct. 6 of last year was down significantly compared to that found earlier with payloads aboard Pioneers III and IV, and the Soviet Mechta.

On the night of August 16-17, a strong geomagnetic storm occured which brought a new surprise. During the first 24 hours of the storm, about two-thirds other radiation in the outer zone was "dumped" on the earth.

Available evidence suggests that the lost radiation consisted largely of low energy electrons.

The outer belt soon recovered, in fact over-recovered, and near the end of the storm, on August 18, the outer belt was found to contain about five times the total radiation found there before the storm.

This abnormally high radiation "fever" persisted for ten days and was followed by a slow return to normal.

As yet, a satisfactory explanation of the "dumping" process in the radiation of the outer belt has not been found.

Tracing the probable dumping paths of this radiation to earth, the Minnesota scientists found that it should show up near the surface of the earth between magnetic latitudes of 52 and 62 degrees, with a peak appearance at 57 or 58 degrees.

On the very night of the magnetic storm, the counters aboard Explorer VI revealed the disappearance of radiation from the outer belt, Dr. Winckler observed an aurora over Minneapolis, at a magnetic latitude of 57 degrees.

The tie-in was inescapable. The observed aurora must have been caused by the jumped electrons from the outer belt.

Such electrons would produce more than the visible evidence of an aurora. They would produce X-rays. Previous observations by balloon had already indicated that such X-rays were often present at relatively low latitudes during times that visible auroras lay far to the north.

The Minnesota data thus gives support to the idea that under certain conditions at least, an aurora may consist of two parts, a visible aurora most likely appearing near the usual auroral zone, and an X-ray aurora, invisible perhaps to the naked eye, lying to the south.

The National Aeronautics and Space Administration picked up the tab for this work, and most likely will do so in the future as the Van Allens, Arnoldys, Hoffmans, Wincklers and others bring new light to bear on the remaining mysteries of the earth-circling radiation belts. Salt Lake Tribune, June 19, 1960.

SPACE FILMS DISCLOSE FAR GREATER RADIATION

BEDFORE, MASS. (UPI) - Film packs exposed at altitudes up to 700 miles in an Atlas nose cone showed far greater radiation than previously discovered, according to the air research and development command.

"Particle tracks indicate that radiation in the (Van Allen) belts discovered by the Explorer satellites far outshadow the flux of primary cosmic radiation as measured by means of skyhook balloons at 20-mile altitudes," the report said.

"Counts in the Atlas-flown emulsions show that for every cosmic ray that penetrated the block, some 44,000 protons of the Van Allen belt were recorded.

"The accompanying photomicrograph of a section of the emulsion smaller than the head of a common pin shows a track produced by a heavy cosmic ray primary immersed in a sea of protons carrying kinetic energies in excess of four billion electron volts and thus capable of penetrating the Atlas nose cone.

"The more energetic of the trapped particles produced 'stars' or nuclear explosions inside the emulsion (film) 500 times more frequently than observed on earlier Aerobee rockets which reached only 100 miles, below the intense trapped radiation of the Van Allen belt." (March 17, 1960. **Deseret News),** Salt Lake City)

APPENDIX II

Deletion from original chapter 4 now chapter 10

EARTH FLOATS IN ELECTRICAL SEA

By Alton L. Blakeslee — Associated Press Science Writer

BOULDER, COLO., Oct. 12, 1957 (AP) - High overhead floats a fantastic electrical sea.

It girdles the earth, reaches at least 150 miles deep. It writhes with storms and savage winds. Powerful electro-jet currents course through it.

It is pulled by tides, pocked by peculiar clouds, bombarded by cosmic rays.

Created by the sun's cruelest rays, this sea is the ionosphere, a vast belt of electrons and electrified atoms or ions. It begins 60 miles up, goes at least 200 miles high.

In sparsest form it apparently reaches thousands of miles into desolate space.

It's a shield between you and a deadly sun.

Were it not there, absorbing the sun x-rays and most power ultraviolet light, life on earth would perish.

Were it not there, you might never hear a radio. Short wave radio communication depends upon bouncing or reflecting radio waves back to earth from this electrical sea.

Strange quirks in the ionspheres sometimes perform magic. Miami police calls are heard in California.

Or a picture from a TV station hundreds of miles away suddenly appears on the sceen.

Exploring this sea is a major activity of the International Geophysical Year, a co-operative 64-nation effort to learn more about our earth, sun and space.

Fingers of radio itself are a prime method of ionsphere exploring. Literally thousands of times a day over the world special radio beams are darting up and bouncing back to measure heights, intensities and other changing peculiarities of the ionosphere.

When the full story is pieced together, scientists hope to answer some puzzles of the high atmosphere and find new or improved ways of putting the ionosphere to human service.

The "radio" fingers and other techniques already have disclosed much of the story, explains Robert W. Knecht, a project leader in sun-earth relationships at the National Bureau of Standards Boulder laboratories.

In reality the ionosphere is our outer atmosphere of ultra-thin air. X-rays and ultraviolet light from the sun rip into molecules of oxygen and nitrogen, tearing out their electrons, electrifying billions times billions of atoms.

Usually the ionosphere has distinct layers.

About 60 miles high is the E layer, then the denser, F-1 region at about 120 miles, the F-2 layer at 200 miles.

The "E" layer reflects low-frequency or long radio waves. Higher frequencies or shorter waves penetrate through it, bounce back from higher layers. Sufficiently high frequencies barrel right on through into space. Usually this is what happens with TV signals.

During IGY, nearly 200 special radio-sounding stations from pole to pole are intently exploring the ionosphere. Each shoots up pulses of radio waves, sweeping through a quick range from long to short waves in 15 seconds, then timing and recording the echoes from different layers.

A few will make continuous recordings of the seething electrical sea.

For the ionosphere, far from being a static shell, changes, minute by minute, hour by hour, season by season. It is a sensitive link between events on the sun and earth, Knecht points out.

Great flares or explosions on the sun sometimes create a much enhanced D layer about 40 to 50 miles up. This absorbs rather than reflects radio waves, producing radio blackouts and interference. Other sunflares have no effect, for reasons not yet understood.

NOTED PHYSICIST ANNOUNCES NEW ENERGY SOURCE

Nickola Tesla, the renowned physicist and inventor who developed the great AC motor, the fundamental principle of radio, and the practical transmission of alternating power foretold the discovery, many years ago of a hitherto unknown source of unlimited energy, "so practical that the machinery to harness it will last 500 years, and so basic that it will undo existing theories."

"They called me crazy, in 1896," said Dr. Tesla, "when I announced the discovery of cosmic rays. Again and again they jeered when I discovered something new and then years later saw that I was right. Now I suppose it will be the same old story when I say I have discovered a hitherto unknown source of energy, unlimited energy, that can be harnessed.

"The initial cost will be relatively big. After that hardly anything and unlimited power for the asking."

Dr. Tesla has given the world the arc lighting system, the Tesla coil and rotating field principle for alternate current and innumerable other electrical devices.

Dr. Tesla did not live to reduce a practical application the discovery he referred to above.

"ASTRONOMY," by Robert H. Baker, Ph.D. Professor of Astronomy, University of Illinois, page 303:

Another problem relates to the apparent lavish expenditure of this radiation. Of all the energy that pours forth from the sun, less than one part in 200 million is intercepted by the planets and their satellites, The remainder spreads through interstellar space with little chance, so far as we know, of being recovered. The suggestion that the sun shines only in the directions of material that can intercept it makes an appeal from the point of view of economy, but appears to have little else to recommend it. It would seem that nature is squandering its resources of energy so prodigally that it must end in bankruptcy; but doubtless we have at present an imperfect account of the situation."

Foundations of the Universe, Luckiesh, General Electric, pp. 41-43 **Astronomy,** Robert H. Baker, Ph.D., page 303

The great success of the atomistic principles as it is involved in the kinetic theory of matter was one of the wonders of the modern scientific age. It is to be expected that there will be found other applications equally fascinating and promising. It is now being pressed further into the service of explaining the structure of matter . . .

When Maxwell (1873) propounded the electro-magnetic theory of light (radiation), his achievement was epochal. The exact manner in which the Radiant Energy traversed space was not known, and the next epochal event was the founding by Planck (1900) of the quantum theory. Here we have the atomistic principle applied to energy instead of being confined to the material of the universe as it had been. In other words, in the quantum theory we have the atomistic idea applied to physical processes. We now have the atom of matter, the atom (electron) of electricity, and the atom (quantum) of action (a product of energy and time). Planck assumed the emission of radiation (from the sun, a lamp filament, etc.) to occur discontinuously. He conceived elements of energy of equal magnitude analogous to the equality of electrons, or atoms of a given element. Radiation or Radiant Energy is emitted of various wave lengths or frequencies which must be taken into account in laws of radiation...now the physicist uses quanta as commonly as he does electrons and atoms and molecules. Bodies are built of molecules, the molecules of atoms, and the atoms of electrons (and protons) etc. Here we see the atomistic principle applied to "material" (matter) and then to electricity. Finally, a physical process — the radiation omitted by the electrons — is divided into quanta. With such pictures of the Universe being considered we may cease to be surprised at anything, but our interest and admiration will grow. Will we ever get to the final foundation?

APPENDIX III
THE INSTANTANEOUS POWER AVAILABLE FROM ELECTRO-MAGNETIC WAVES

by Richard E. Maxwell, Ph.D.

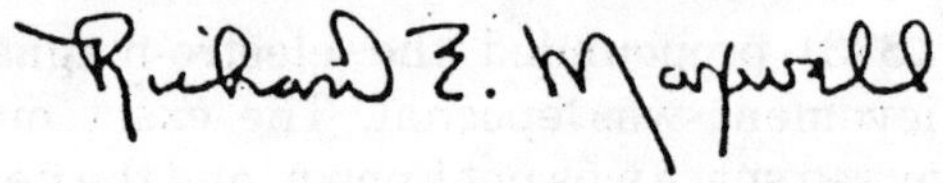

Introduction

The objective of this study was to show from a theoretical point of view the magnitude of the instantaneous power which is available at the Earth's surface. The source of these electromagnetic waves (EMW) was taken to be a single solar source which generates the EMW as a result of nuclear reactions. It was assumed that the EMW reached the Earth essentially as planes and that there was no reflection of these waves by objects on the Earth's surface. A number of solar sources could have been considered as could have reflections of the EMW; however, this type of analysis is difficult and complicated. Hence, the approach taken for this study was to consider a realistic, but simplistic setting.

Conclusions

1. If the EMW emanate from a single source, and if they are absorbed into a cube of volume d^3, then the instantaneous power available is given by

$$P = d^2 E_x H_y$$

where

E_x is the magnitude of the electric field

H_y is the magnitude of the magnetic field

2. If $H_y = 39.79$ amps/meter (equivalent to a B field of 5×10^{-5} teslas), $E_x = 100$ volts/meter, and if d = 1 meter, then the power, P = 3979 watts. In order to obtain 50,000 watts, a 1 cm diameter antenna of length 400 meters would be required. This implies that the Radiant Energy device is probably **not** detecting only electro-magnetic energy.

3. The product of the field intensities, $E_x H_y$ should be a periodic function with a period of 24 hours. This would imply that the power which is available would be a maximum at mid-day and essentially zero at mid-night, and because of the line of sight of the EMW, more power would probably be available at the equator than at the poles.

4. In the book, **The Sea of Energy in Which the Earth Floats,** it was reported that energy was available for conversion and capture at night where

**Dr. T.C. Jones, University of Arizona, Tucson, Arizona*

the magnitude of this energy was not as large as by day. From the latter conclusion, the energy expected at mid-night is essentially zero, and this conclusion contradicts reported experimental results. This contradiction implies there are multiple sources of EMW, and it would be interesting to consider the moon as a solar reflecting body. It seems to me, however, that the area of the moon would not be large enough to reflect the power as detected at night and reported in the book. Hence, one must conclude there are multiple sources of EMW.

Analysis

Reference: **Fields and Waves in Communication electronics,** Ramo, Whinnery, Van Duzen, 1967, Wiley Publications, p. 243.

Energy flow into a closed surface, S, is given by

$$\text{Power} = W = -\int_s \vec{E} \times \vec{H} \cdot dS$$

Suppose that current flow and charge in this region are zero in a medium which is homogeneous, isotropic, linear, and with zero conductivity.

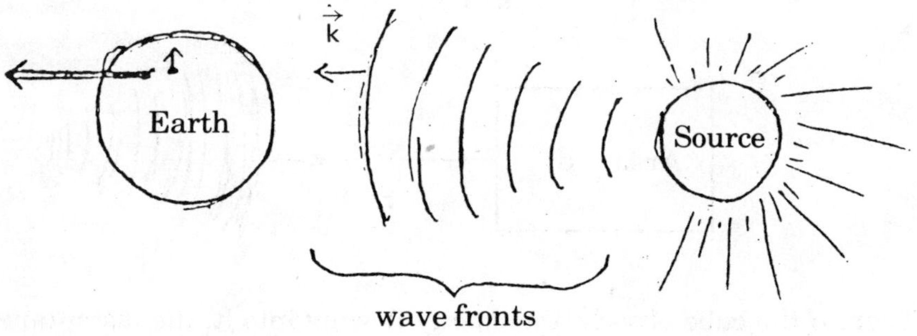

$$F = F_x\vec{i} + F_y\vec{j} + F_z k$$

$\vec{k}$ is the direction of propogation $-z$ direction

$$\vec{\nabla}\cdot\vec{\nabla}\vec{E} = \left(\frac{\partial^2}{\partial x^2} + \frac{\partial^2}{\partial y^2} + \frac{\partial^2}{\partial z^2}\right)\cdot(E_x\vec{i} - E_y\vec{j} - E_z\vec{k}) = \mu\epsilon\frac{\partial^2 E}{\partial t^2}$$

Assuming $\frac{\partial}{\partial x} = \frac{\partial}{\partial y} = 0$ implies a uniform planar wave which implies

$$\frac{\partial^2 E_x}{\partial z^2}\vec{i} + \frac{\partial^2 E_y}{\partial z^2}\vec{j} + \frac{\partial^2 E_z}{\partial z^2}\vec{k} = \mu\epsilon\left\{\frac{\partial^2 E_x}{\partial t^2}\vec{i} - \frac{\partial^2 E_y}{\partial t^2}\vec{j} - \frac{\partial^2 E_z}{\partial t^2}\vec{k}\right\}$$

Taking $\vec{i}$ to be in the direction of $E = E_y = E_z = 0$;

hence, we have $\frac{\partial^2 E_x}{\partial z^2} = \mu\epsilon\frac{\partial^2 E_x}{\partial t^2}$ implies wave velocity $= \frac{1}{\sqrt{\mu\epsilon}}$

where $\vec{E} = E_x(z,t)\,\vec{i}$

$$\vec{\nabla} \times \vec{E} = -\frac{\partial B}{\partial t} \text{ where } \vec{B} = \mu\vec{H} - \mu(H_x\vec{i} + H_y\vec{j} + H_z\vec{k})$$

$$\vec{\nabla} \times E = \begin{vmatrix} \vec{i} & \vec{i} & \vec{k} \\ \frac{\partial}{\partial x} & \frac{\partial}{\partial y} & \frac{\partial}{\partial z} \\ E_x & E_y & E_z \end{vmatrix} = o\vec{i} + \frac{\partial E_x}{\partial z}\vec{j} + o\vec{k} \text{ implies}$$

$$o\vec{i} + \frac{\partial E_x}{\partial z}\vec{j} + o\vec{k} = -\mu\left(\frac{\partial H_x}{\partial t}\vec{i} + \frac{\partial H_y}{\partial t}\vec{j} + \frac{\partial H_z}{\partial t}\vec{k}\right)$$

Setting $H_x = H_z = O$ implies

$$\text{Power} = -\int_s(E_x\vec{i} \times H_y\vec{j})\cdot dS = -\int_s(E_xH_y\vec{k}).dS$$

Assume waves from only a single direction, and consider the cube of side d—one side of which is orthogonal to $\vec{k}$ i.e.,

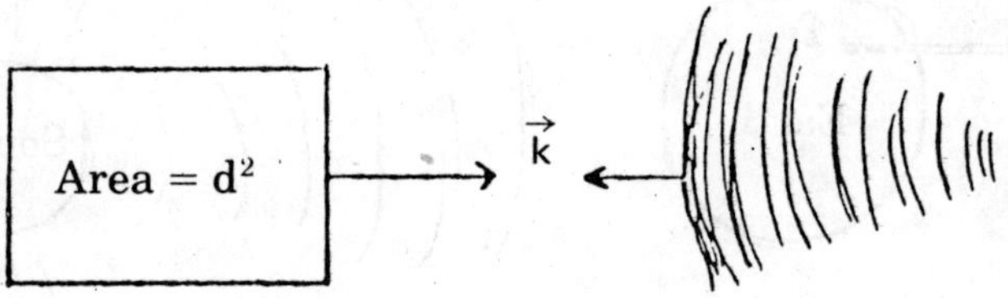

Then, if the cube absorbs the energy flowing into it, the instantaneous power absorbed is given by

$$\text{Power} = -\int_A E_xH_y dA = -AE_xH_y = -d^2E_xH_y$$

More generally,

$$\text{Power} = -A\left|\vec{E} \times \vec{H}\right|$$

The time dependent function $E_xH_y = f(t)$ should be periodic with period = 24 hours. The half-range Fourier series expansion of this function is given by

$$E_xH_y = \tfrac{1}{2}a_o + \sum_{k=1}^{\infty} a_k \cos\left(\frac{k\pi t}{p}\right) \qquad \text{p=period}$$

where $$a_k = \frac{2}{P}\int_o^p f(t)\cos\left(\frac{k\pi t}{p}\right)dt$$

This implies Power $= -A\left\{ \tfrac{1}{2}a_o + \sum_{k=1}^{\infty} a_k \cos\frac{k\pi t}{p} \right\}$

where $a_k = \frac{2}{P}\int_o^p E_x(t)H_y(t)\cos(\frac{k\pi t}{p})dt$

APPENDIX IV

ON THE SUBJECT OF RADIANT ENERGY

by C. Warren Simmonds, Ph.D.

Many forms of Radiant Energy are known to be arriving at the Earth's surface from outer space. The form of Radiant Energy most familiar to everyone is the heat and light rays from the sun. These rays have been studied and measured by scientists for many years and their properties are well known and documented. Other forms of Radiant Energy are less familiar: Cosmic rays; gamma rays; galactic noise; radio frequency emanations from the stars; gravitational waves; and other forms yet unknown and undetected.

With all the Radiant Energy traveling throughout the universe, presumably in all directions, it is not an exaggeration to say that the Earth itself is immersed in energy, or as expressed by Dr. T. Henry Moray of the Research Institute, Inc. in Salt Lake City, Utah, the Earth is "floating in a sea of energy".

Now with the Earth surrounded on all sides by energy it doesn't make sense that mankind should ever suffer from an energy shortage, provided (1) He can use the energy directly, as in the case of sunlight, or (2) He can **convert** the energy to a useable form. An intriguing question is: Can a form of this Radiant Energy, known or unknown, be detected on Earth and converted to a useful form of energy, such as electrical power? The following calculations will show that Radiant Energy, regardless of its exact form, arriving at the Earth's surface at the usual speed of light, can result in significant amounts of power (50 KW, for example) at the Earth's surface for very low energy densities (200 μ joules per cubic meter, or less) in space.

Consider an ideal detector for Radiant Energy at the Earth's surface, having an effective cross-sectional area, A, set perpendicular to the direction of the arriving Radiant Energy, as shown in Fig. 1. The Radiant Energy is shown arriving at the velocity of light, c, and the density in space is taken as ρ joules per cubic meter. An imaginary column is shown above the area. For each second of time, such a column of length equal to c meters will pass into the ideal detector, carrying a total energy of

$$cA \times \rho = pcA \text{ joules}$$

Since this number of joules is received every second, the detected power, P, is simply

$$P = \rho cA \frac{\text{joules}}{\text{sec}}, \text{ or watts} \quad (1)$$

Using (1), we can determine the density P in space necessary to receive 50 KW of power using a detector whose effective area, A, is only one square meter.

Substituting into (1), we have

$$50 \times 10^3 = \rho \times 3 \times 10^8 \times 1$$

or,

(2) $$\rho = \frac{50 \times 10^3}{3 \times 10^8} = 166 \times 10^{-6} = 166 \text{ microjoules per cubic meter}$$

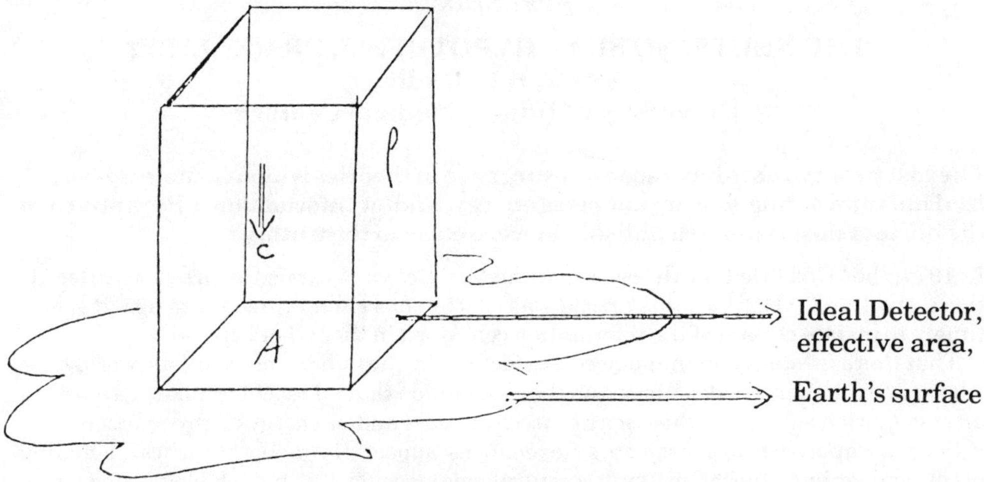

Fig. 1 - Radiant Energy arriving at the Earth's surface

Now, 166 microjoules is a very small amount of energy. To get an idea of how small it really is, it represents the energy stored in a 3.3 μfd capacitor charged to 10 volts. This same energy would only raise the temperature of one cubic centimeter of water

(3) $$166 \times 10^{-6} \text{ joules} \times \frac{\text{gm-calorie}}{4.186 \text{ joules}} = 39.6 \times 10^{-6} \text{ degrees Centigrade}$$

which is approximately 40 millionths of one degree centigrade temperature rise.

So, the actual energy density of Radiant Energy as it travels through space can be very small and yet result in what we would call significant amounts of power. The 50 KW received by the detector having an effective area of one square meter in our example corresponds to

(4) $$50{,}000 \text{ watts} \times \frac{\text{HP}}{746 \text{ watts}} = 67 \text{ Horsepower}$$

which is sufficient to adequately power an automobile.

Although the exact nature and form of Radiant Energy is not fully understood, it has already been successful detected. In 1939, Dr. T. Henry Moray, using a specially built apparatus, detected 50 KW of useable electrical power in experiments witnessed by certain distinguished scientists.

APPENDIX V

THE NEUTRINO SEA—HYPOTHESIS, OR REALITY?

by Dr. H.C. Dudler
University of Illinois Medical Center

The last twenty years have seen a resurgence of theories which assume a generalized medium acting to transmit mass energy, and/or information. Perhaps then the ancient postulates of a phlogiston were close to the truth.

In 1974, the "Unsettled Earth" issue of *Industrial Research* carried an article entitled "Is there an ether?" (1). Since this article was written more than three years ago, it seems timely to update certain of the information contained in the "ether" article.

That there is interest in the subject is evidenced by the number of scientists, worldwide, who are developing "models" based on the assumption that all reactions take place within an energy-rich substrate which may contribute mass and/or energy to any reaction.

Such an approach to mass/energy exchanges appears in many branches, including plant and animal biochemistry, structural engineering (2), microbiology, and quite extensively in astrophysics.

One aspect of the climate of opinion which dominates the physical sciences of 1977 results from a series of experiments often referred to as the Michelson-Morley studies, carried on before 1890. Using more modern apparatus, other investigators essentially confirmed these earlier M-M studies, concluding that no subquantic medium is necessary for that transmission of electromagnetic radiation.

But all of these experimental designs failed to consider the Earth's motion about the center of our galaxy, 220 km/sec, or the orientation of the Earth with respect to this motion. From the astronomic information now available, it appears that the flow of the "ether" that was being sought was nearly normal to the plane of the apparatus, thus insuring the "null" results obtained (3).

A philosophical tenet long taught in all beginning physical science classes holds that absolute motion cannot be demonstrated. This concept originated with Isaac Newton (*ca* 1700); it provided some of the basic assumptions of Einstein's theoretical studies (1905-1915).

If one analyzes the experimental designs of M-M-type studies, it will be noted that what was being sought was a primary or absolute frame of reference whereby the lateral motion of the Earth might be demonstrated without reference to any other body.

This demonstration first was accomplished by Conklin (1969) (4) on study of the diurnal variation of the 3.5-cm radiation flux which bathes the Earth. The Earth's motion was estimated to be 160 km/sec in a direction defined in local coordinates as right ascension 13 hr: declination 32 deg.

Recently, Corey and Wilkinson (5) have shown that there is a generalized flux of high frequency electromagnetic radiation (19 GHz) in which the Earth is immersed. Their observations indicate a velocity of 330 km/sec for the Earth with respect to this "soup." These studies utilized a balloon-borne radiometer, with the lower frequencies, as used by Conklin, tuned out.

Smoot, Gorenstein, and Muller (6) have studied a still higher-frequency background radiation flux (33 GHz) at a height of 20 km; they have demonstrated motion of the Earth with respect to this flux, 320 km/sec, toward a point in the sky located at 11 hr right ascension: declination 6 deg.

Results of the various studies may vary. Yet a ball-park figure begins to emerge. It is an unescapable fact that absolute motion certainly has been demonstrated. The rush of new techniques, based on new apparatus, extending the lab bench high into space, has rendered some of our most hallowed theoretical dogma untenable. Time (and Physics) marches on!

As a result of the M-M-type studies, combined with the theoretical/mathematical approach to nuclear science, fashionable during the past 40 years, there has evolved an assumption that a vacuum (i.e., empty space) is essentially inert: mass-free and energy-free. This assumption recently has led to an unusual state of affairs in the fields of chemistry and physics.

Two investigators, working independently, have been nominated jointly for a Nobel Prize. But the nature of their work has largely escaped the attention of the U.S. scientific community.

Kervran of France (7) reports that oat seedlings have been shown to convert potassium to calcium during the germination phase, with an increase of Ca in the range of 100 to 163%.

Komaki of Japan (8) reports that eight strains of micro-organisms grown in K-deficient culture media increase the total K by converting Ca to K.

These studies appear to have been confirmed by other investigators. A theoretical basis for Kervran's result has been discussed by de Beauregard (9), who postulates that K, by the addition of a proton (H^+), is converted to Ca. Reasoning on the basis of presently accepted nuclear theory, this reaction would require energy of 10 MeV.

The generalized neutrino sea is considered to be the source of this energy thru the action of "neutral currents" and "weak forces." These mechanisms are those now in general use of particle physicists to explain the interaction of a host of other subnuclear entities.

During the past two decades, there has been a resurgence of theories which have in common, as one of their basic assumptions, the existence of some type of generalized medium which acts to transmit mass, energy, and/or information. To better understand the present, and to obtain an overview of the importance of this subject, let us examine one aspect of scientific history.

The necessity of the same type of generalized medium was postulated by Isaac Newton (*Opticks,* Book 3, Part 1, Query 18, 1704):

"And do not hot Bodies communicate their Heat to contiguous cold ones, by the Vibrations of this Medium propagated from them into the cold ones? And is not this Medium exceedingly more rare and subtile than the Air, and exceedingly more elastick and active and doth is not readilly pervade all Bodies? And is it not (by its elastick force) expanded through all the Heavens?"

This concept had altered but little into the first decade of the 20th Century (Textbook of Physics, Edit. A.W. Duff, 1912, 3rd Edit, p. 565.):

"The Ether.—To account for the transmission of waves through space containing no ordinary matter it seems necessary to assume the existence of a universal medium filling all space and even interpenetrating matter itself, as shown by the existence of transparent substances. That this medium can react on matter is shown by the fact that radiant energy is transmitted from ether to matter in the case of absorption, and from matter to ether in the case of emission of radiation by material sources."

The version which now so rigidly defines the climate of opinion of 1977 physics had its beginning in the early 1930s (Textbook of Physics, Edit. A.W. Duff, 1932 7th Edit, p. 556):

"In recent years doubt as to the necessity for assuming the existence of an ether has been expressed by some who believe that it is sufficient to attribute the power of transmitting radiation to space itself. We cannot discuss the question here, but pending the settlement of the controversy it seems wise to continue the use of the word ether as at least denoting the power of space, vacant or occupied by matter, to transmit radiation.

"The 'unreality' of the ether brought out so emphatically by the theory of relativity has recently become even more pronounced, in connection with the new wave mechanics. It seems very doubtful, at the present time, whether light waves have any real existence at all. It is, however, an undisputed experimental fact that light exhibits the properties of waves, and in the subsequent paragraphs we shall so consider it, without further regard to the ultimate nature of the wave motion."

As a result of the experimental proof by Reines and Cowan (1953) of the neutrino as a discrete particle, I posed the following question in formal lectures:

"According to recent laboratory studies, the neutrino seems to have a half-life near infinity, penetration of lead with a half-thickness measured in light-years, and at least one quantum number, spin."

Question: "What has happened to all the neutrinos produced by beta and meson decay thoughout all times?" (1957).

Having been indoctrinated early in my scientific training as to the logical necessity of some sort of "ether," I concluded that a neutrino flux, generated by the myriad of nearly randomly distributed stars, appeared to be the long sought characterization of Newton's "Aetherial Medium." I wrote of this (10) and presented papers before the American Physical Society (Bulletin APS. 7,568,609, 1962):

"DeBroglie (1959) has suggested that certain phenomena arise from the interaction of particles with a 'subquantic' medium, which escape our observation, is in random motion, and is everywhere present in what we call 'empty space.' Chiu suggests (1962) that we exist in a flux of the order 1 x 10 10^{11} neutrinos cm^2/sec. Thus, it seems that both theoretical and experimental physics are moving rapidly toward a concept of quantized fields, or more properly a particulate 'ether.'

"Observed net energy of nuclear events may be due to the exchange of mass and kinetic energy between two classes of particulate matter, i.e., interaction as at the interface of a two-phase system:

$$\begin{array}{lcl} (M_a) + (E_a) & \nearrow\!\!\!\searrow & w(M_a) + x(E_a) + (1-y)(M_b) + (1-z)(E_b) \\ \hline (M_b) + (E_b) & \searrow\!\!\!\nearrow & y(M_b) + z(E_b) + (1-w)(M_a) + (1-x)(E_a) \end{array}$$

where (M_a) = total mass of the neutrinos entering into the reaction; (E_a) = total energy of these neutrinos; (M_b) = total initial mass of the nucleons; and (E_b) = total energy of the nucleons entering into the reaction.

"Mass and energy would then be exchanged by the two systems, momentum being conserved, with no interconversion of mass and energy."

In addition to those names given above, others wrote of "hidden variables," "neutrino sea," etc.,—1962 to 1965—Bohn, Vigier, Ponticorvo, S. Weinberg, de Silva, Lochak, et al. The volume of literature has increased steadily; recent review articles now cite hundreds of pertinent papers.

As a result of the data and development of concepts in astrophysics, for cosmological considerations, the neutrino sea has been defined as an energy-rich, particulate, generalized medium consisting of electron neutrinos, $\sim 10^{12}/cm^3$, energy density estimates ranging from $10^8 eV/cm^3$ to as high as $10^{10} eV/cm^3$(11).

All this plus muon neutrinos of mass ~0.6 MeV, which appear to be the counterpart of the uncharged electron.

There is a practical aspect to these finer points of nuclear theory. That is the implicit assumption that currently accepted theories define all parameters of all nuclear reactions.

Such a state of mind is illogical, for it flies in the face of all previous experience in the sciences. It has been repeatedly demonstrated that new apparatus and new techniques produce new and unexpected experimentally demonstrated facts and relationships, thru which all theories are eventually rendered untenable.

Quo vadis

The unpleasant aspect of inducing vast nuclear accidents has been examined previously—just prior to the first fission explosions in 1945 (12), and more recently in discussions of the propagation of fusion reactions in the sea (13). In short, what is the probability that man can inadvertently induce catastrophic nuclear events simply as the result of lack of basic information; or as the result of not using the information now available?

There is considerable opposition to even the cursory examination of this approach to nuclear science (14, 15, 16). The reason is clear from the text of the 1932 quotation above. If there exists a subquantic medium with which it is possible to interact, then the bulk of modern nuclear theory is untenable.

Several Nobel prize winners who are fearful of our present headlong rush into nuclear technology, have pointed out that those who have spent their lives developing this field now see the usefulness of their work being questioned. The nuclear leaders of the past 30 years react quite humanly, for they are most reluctant to accept this turn of events.

Like it or not, the fact that methods of establishing absolute motion have been found—that experimental evidence is at hand — indicates that we do exist in an energy-rich subquatic medium — that we can and are interacting with this medium.

All of this information indicates the weakness of present official projections of limits of nuclear risks. Like it or not, probing questions are indeed, being asked, worldwide. And no amount of face-saving maneuvers will be effective in stifling the subject.

Scientific revolutions can be defined as periods of time in which man is forced to reorient his basic viewpoint with respect to the workings of the physical universe in which we are observers and passengers, but not pilots. These revolutions are the direct result of the development of improved accuracy in apparatus with which to extend man's five senses.

New concepts evolve out of the attempts to correlate and understand the data which new apparatus produces. Thus, thru evolutionary processes, new paradigms replace old viewpoints, and man is better able to understand the workings of the macro-cosmos and the micro-cosmos. He nibbles away at his vast store of ignorance, bit by bit, generation after generation.

Modern man has generated three scientific revolutions, each of which required him to make radical changes in his understandings of cause and effect. The studies of Copernicus and Kepler, when integrated into the climate of the discovery guiding Newton and his contemporaries, (*ca* 1675) shattered and ego of the ancients, for the Earth was not at the center of creation, it was but one of several planets orbiting our Sun.

Earth, a speck

This revolution has been extended by the 200-in. telescope, by photo and electron imaging techniques, and recently by the use of interplanetary platforms for carrying all manner of sophisticated electronic extensions of our eyes. The Earth is now but a speck on the outer rim of one galaxy, hurtling thru intergalactic space at a velocity of 200 to 300 km/sec. We don't know where we have been nor where we are going. Neither did the ancients.

The 19th Century saw the clarification of the nature of air; the combustion process was defined; heat was shown to result from a kinetic process involving moving atoms, rather than being a fluid called caloric or phlogiston. Of such was the second scientific revolution composed.

Beginning in 1895, the third revolution required man to alter radically his concepts regarding the nature of the building blocks which were required to construct all that which man could observe — in fact, that which constituted man himself.

The elements which heretofore were thought to be immutable, were sometimes rather easily altered. In fact, some were changing into other atoms by some process so mysterious that it was thought to be spontaneous, occurring without prior cause.

This third revolution produced such an avalanche of new data that the present years are being called the "Information Explosion."

The fourth scientific revolution is an out-growth of the new data, new apparatus, new concepts. Man now has begun to study and to define the particulate substrate which permeates all space and all matter.

The findings from these studies require man again to reorient his thinking and attitudes with respect to the Universe about us. But because of inertia, because of a love affair with the *status quo,* those indoctrinated in science 1977 may have to be forced to move ahead into the 21st century of sub-electron and subquantic physics.

References

1. Dudley, H. C., Is there an ether? *Industrial Research.* Nov. 15, 1974, p. 97.
2. Grimer, F.J., and Hewitt, R.E., Structure, Solid Mechanics & Engineering Design, John Wiley & Sons. London, 1971, p. 681 and 863.
3. Dudley, H.C., *Bull. Atomic Scientists,* Jan. 1975, p. 47.
4. Conklin, E., *Nature,* 222, 1969, p. 971.
5. Corey B.E., and Wilkenson, D.T., cited in *Science News,* July 3, 1976, p. 10.
6. Smoot, G., Gorenstein, M.V., and Muller, R. A., cited in *Science News,* July 16, 1977, p. 44.
7. Kervran, C.L., *Prev. en Biologie de Transmutation a Faible Energie,* Maloinx, Paris, 1975.
8. Komaki. H., *Rev. de Pathologie Comparee,* 67, 1967, p. 213; 69, 1969, p. 29.
9. de Beauregard, O. Costa, Proceed. 3rd. Int. Cong. Psych. (Tokyo), June 1977, p. 168.
10. Dudley, H.C., *New Principles in Quantum Mechanics,* Exposition - University Press, New York, 1959.
11. de Graaf, T., Cosmic Background of Low-Energy Neutrinos, *Astronomy & Astrophysics.* 5, 1970, p. 335.
12. Konopinski, E.J., Marvin, C., and Teller, E., Ignition of the Atmosphere with Nuclear Bombs, Los Alamos Report La - 602 1945. Classified Secret. Declass. Feb. 1973.
13. Dudley, H.C., Ultimate Catastrophe, *Bull. Atomic Scientists,* Nov. 1975, p. 21; June 1976, p. 38.
14. Bethe, H., *Bull. Atomic Scientists,* June 1976, p. 36.
15. Feld, B.T., *Bull. Atomic Scientists,* June 1976, p. 38.
16. Dudley, H.C., Morality of Nuclear Planning? Kronos Press, Glassboro, NJ, 1976.

APPENDIX VI

Article reprinted from **Lettere Al Nuovo Cimento, Vol. 13, N. 1, May 3, 1975; with permission.**

LETTERE AL NUOVO CIMENTO VOL. 13, N. 1 **3 Maggio 1975**

Neutrino Structure of the Ether.

M. RUDERFER
Dimensions, Inc. - Hempstead, N.Y.

(ricevuto il 10 Gennaio 1975)

Although existence of an ether is widely rejected, this is not a precise interpretation of relativity is stated, the concept of the aether as a *substance* is thereby removed from the physical theories. For there is no point in discussing a state of rest or of motion relative to the aether when these quantities cannot, in principle, be observed experimentally...It should no longer be regarded as a substance but simply as *the totality of those physical quantities which are to be associated with matter-free space*. In this wider sense there does, of course, exist an aether; only one has to bear in mind that it does not possess any mechanial properties."

Accordingly, the widespread rejection of an ether applies to a Galilean ether, but not to a Lorentz-invariant ether since its description of observation is necessarily identical to that of relativity. The Lorentz ether has furthermore been demonstrated to be inseparable from relativity (2), has potential advantages for physical theory (3) and is compatible with one-way light tests (4). The most direct verification of the validity of a Lorentz ether, and one which is independent of relativity, is found in the theoretical works of IVES. Building on the earlier theories of FITZGERALD, LORENTZ and LARMOR, he produced a classical derivation of the Lorentz transformations (5), showed the consistency of the Lorentz ether with experiment (6) and extended Lorentz-ether theory to gravitational phenomena (7) by a classical derivation of the Schwarzschild metric, hence to an identical explanation of the three famous tests of relativity. He further showed (8) that the gravitational equivalence of the ether and relativity theories results from the proper correction of Newtonian gravitation, which has an infinite speed

(1) W. PAULI: *Theory of Relativity,* English translation by G. FIELD (New York, N.Y., 1958, p. 4. Italics are in the original.

(2) G. BUILDER: *Austral. Journ. Phys.,* 11, 279 (1958); W. RINDLER: *Amer. Journn. Phys.,* 38, 1111 (1970).

(3) M. RUDERFER: *Lett. Nuovo Cineato,* 3, 658 (1970).

(4) M. RUDERFER: *Amer. Journ. Phys.,* in press. p 3 279 (1975)

(5) H.E. IVES: *Phil. Mag.,* 36, 392 (1945).

(6) H.E. IVES: *Journ. Opt. Soc. Amer.,* numerous papers (1937-1952).

(7) H.E. IVES: *Journ. Opt. Soc. Amer.,* 38, 413 (1948); 29, 183 (1939).

(8) H.E. IVES: *Phys. Rer.,* 72229 (1947).

of propagation, for propagation at the speed of light. As a result, any Lorentz-ether theory of gravitation which provides the static Newtonian gravitational force is necessarily equivalent observationally to general relativity (9).

The admissibility of the Lorentz ether focuses attention on its properties. This ether is not directly observable by any known dynamic or electrodynamic means yet it is associated with the physical attributes of matter-free space, i.e. the vacuum, and therefore has observable energy properties. These include the macroscopic (general relativistic) properties of space-time, which govern the motion of matter and light, and the microscopic (quantum electrodynamic) interactions, as is evident in the Lamb shift and the anomalous gyromagnetic ratios. If the Lorentz ether (≡ the vacuum) is not to be regarded as a substance, what is responsible for its physical attributes?

Knowledge of 'substance' is acquired by direct observation. However, observability is not a constant, but is a function of time because man's ability to measure has been changing rapidly in recent history. What is accepted as a 'substance' may consequently vary with time. Today the various known systems of matter constitute a hierarchy which comprises the observable Universe, galactic clusters, galaxies, stellar systems, ponderable bodies, molecules, atoms and elementary particles; a few centuries ago we could identify with certainty only one level of this hierarchy — ponderable bodies. Furthermore, there is no substantial evidence that the present hierarchy is complete. Each level of the hierarchy also contains unique forms of energy transfer; if Nature's hierarchy extends beyond its present limits, there must exist unknown forms of energy transfer in the Universe which are now unobservable. This is not unlikely because our still accelerating technology presages further expansion of the presently known hierarchy: We are now avidly seeking the structure of elementary particles and our estimate of the size of the Universe has been increasing logarithmically over the last century (10). The probability that the known hierarchy actually extends further then suggests a feasible definition of the ether: It is the totality of all the presently unobservable forms of energy that may exist in the microcosm.

The least detectable class of stable particles in the known hierarchy of matter — the neutrinos — may now be considered as the most prominent component of such an ether. It is presently believed that a vast sea of low-energy neutrinos (and antineutrinos) permeates the Universe; no satisfactory upper limit to the power flux in this sea has yet been accepted (11). Three properties immediately result which are consonant with essential properties of an ether: i) the energy in the neutrino sea is presently unobservable due to the lack of any known means for detecting low-energy neutrinos; ii) the total unobservable energy in the neutrino sea may far exceed the total energy contained in matter, i.e. the vacuum is then a plenum rather than a void; iii) since neutrinos are restricted to speed *c*, the neutrino sea is necessarily Lorentz invariant.

The velocity spectrum of the neutrino sea may be more precisely described in terms of the measured upper limit Δm to the rest mass of neutrinos. Thus

$$m_\nu = \Delta m/(1 - v^2/c^2)^{\frac{1}{2}}, \qquad (1)$$

where $M_\nu C^2$ is the total energy of the neutrino and V is its speed. Letting V= $c - \Delta c$

(9) The continued neglect of Ives' gravitational theory is unwarranted in view of current interest in comparative evaluation of all viable theories, e.g. C. M. Will: *Sci. Amer.*, 231,24 (Nov. 1974); *Phys. Today*, 25, 23 (Oct. 1972).
(10) G. DE VAUCOULEURS: *Science*, 167, 1203 (1970).
(11) M. RUDERFER: *Lett. Nuovo Cimento*, 5, 80 (1972); 10, 826 (1974).

and substituting in eq. (1), we get

(2) $$\Delta c/c = 1 - (1 - \Delta m^2/m_\nu^2)^{1/2}.$$

If $\Delta m/m_\nu \leqslant 2 \cdot 10^{-10}$ for 50 MeV antincutrinos, as recently suggested by LANDE et al. (12), assuming their detected antincutrino pulses originated from a collapsing star, then $\Delta c/c < 2 \cdot 10^{-20}$. If the composite photon theory based on the phasor structure of the neutrinos (13, 14) is assumed, the measured upper limit to the rest mass of the photon applies to propagation of neutrinos with velocity *c* over the same spectral range as has been verified for photons.

A fourth property generally attributed to the ether is an ability to transmit light. This is provided by the neutrino sea in view of a recent analysis by HONIG (15) in combination with the properties of phasor neutrinos. HONIG initially assumed an old ether model consisting of "two oppositely charged continuous superposable fluids" and proceeded with the aid of a toroidal transformation to harmonize the Hertzian description of radiation and planck's quantum restriction. The required polarized fluid is provided by a phasor-neutrino sea in the following way: Phasors are rotating vectors introduced by STEINMETZ over 70 years ago and were subsequently applied extensively in macrosopic applications. They are represented analytically by

(3) $$\psi = \psi_0 \exp[i\theta] = \psi_0 \exp[i\omega t],$$

where ψ_0 is the constant electric of magnetic amplitude, $\theta (\omega = d\theta/dt)$ is rotational angle (speed) in the complex plane and t is time. If such phasors are applied directly t microscopic forms of radiation then in the reference frame of a massless particle and by differentiation of (3) and substitution from E = hw, where E is total photon energy:

(4) $$i\hbar\, \partial\psi/\partial t = -E\psi.$$

For a system with constant total energy, this is identical in form to the conventional Schrödinger wave function except for a difference in sign, which derives from a difference in microscopic and macroscopic conventions. e.g. as for current flow. However, the direct identification of ψ in (4) with an electric or magnetic phasor leads to the feasibility of a composite photon (13, 14) and electron (16) and to observable astrophysical predictions (14, 17). These phasor fields have been physically described by long dipoles but the properties of a neutrino are generally satisfied by one rotating electric dipole moving with velocity *c* (14). BONNOR (18) has demonstrated the consistency with the Maxwell-Einstein equations of dipoles moving with velocity *c*. When such dipoles are applied to the neutrinos of

(12) K. LANDE., G. BOZOKI, W. FRATI, C.K. LEE., E. FUNYVE and O. SAAVEDRA: *Nature,* 251, 485 (1974).

(13) M. RUDERFER: *I.E.E. Journ. Electron. Ponw,* 11, 164 (1965); *Amer. Journ. Phys.* 39, 16 (1971).

(14) M. RUDERFER: NAPS Document No. 01930 (1972). Obtainable from Microfiche Publications, 440 Park Ave. S., New York, N.Y., 10016, by remitting $1.50 (7.55) for microficho (photocopy).

(15) W.M. HONIG: *Found. Phys.,* 4, 367 (1974).

(16) M. RUDERFER: *I.E.E. Journ. Electron. Pow.,* 18, 370 (1972).

(17) M. RUDERFER: *Lett. Nuovo Cimento,* 10, 303 (1974).

(18) W.B. BONNOR: *Intern. Journ. Theor. Phys.,* 3, 257 (1970).

the universal sea, we obtain a vacuum filled with submicroscopic positive and negative charges, i.e. a dually charged fluid as required by HONIG. Although the equivalent neutrino "fluid" is not continuous, neither is any physical fluid. For negligible discontinuity with respect to Honig's mechanism, the ability of the neutrino sea to propagate electromagnetic radiation then follows.

A fifth property generally attributed to the ether is an ability to transmit gravitational forces. Each neutrino dipole may be expressed in terms of finite electric and magnetic susceptibilities (14) which, when averaged over all neutrinos in any volume, provide a physical origin for the permittivity and permeability of the vacuum. The method of Wilson (19) then permits an electromagnetic approach to gravitation equivalent to general relativity. Another equivalent, but more heuristic, approach is provided by the 200-year-old concept of LESAGE which is based on the attractive force obtained between masses situated in a sea of moving microscopic particles which pervade and interact slightly with matter. A modern version (20) may be based on the verified V—A neutrino-electron scattering interaction, or the more intense resonant scattering predicted to account for a variety of astrophysical phenomena (14, 17). If an isotropic neutrino power flux S impinges on a spherically symmetric mass M and there is a fractional transfer of power flux ϵ_m to M by the accumulated scattering interactions, the radial power flux component leaving M has a lower value $(1 - \epsilon_m)S$ which, by the geometry, must be spherically symmetric. There is then a differential flux $\epsilon_m S$ directed toward its center which, from the geometry, must decrease as $1/r^2$, where r is distance from the center of M A test mass m at large r then experiences an unbalanced power flux in the direction of M, hence an unbalanced static force of

$$F \propto [\epsilon_M S - (1 - \epsilon_m)\epsilon_M S]/r^2 = \epsilon_m \epsilon_M S/r^2 \propto mM/r^2 \tag{5}$$

since the E's, from the Compton-effect kinematics, must be dependent only upon total mass and energy of the interacting particles when averaged over all directions. This force is undistinguishable from gravitation and accounts fully for the Newtronian static force if the E's and S have appropriate values. Correction for the finite velocity of neutrinos after IVES, as discussed above, provides the observational equivalent to the general-relativistic description.

If an isolated mass is stationary in an isotropic sea, scattering interactions averaged over all directions produce no net motion of any of its particles, the mass remains at rest and there is no heat production. However, to the extent that there is randomness in the neutrino flux, which must be small if the neutrino number density is high, each particle experiences small random variations in interaction. For electrons this provides a potential mechanism to account for $1/f$ electrical noise (f = noise frequency) and, for more massive particles, a potential involvement in the origin of radioactive decay (21). Heat production per unit mass due to random motion is necessarily negligible, but accumulation in large masses may become significant. If so, this should be masked by nuclear heat generation in stars and may become first noticeable in large masses that have not achieved nuclear burning, e.g. in regard to excess heat production of Jupiter.

If an isolated mass is moving uniformly in the neutrino sea, the accumulated effects of

(19) H.A. WILSON: *Phys. Rev.*, 17, 54 (1921).

(20) M. RUDERFER: NAPS Document No. 02052 (1973), Noto 54. Obtainable from Microfiche Publications, 440 Park Ave. S., New York, N.Y., 10016, by remitting $1.50 (12.95) for microfiocho (photocopy).

(21) Rof. (10), Soct. 16.

all scattering components normal to the motion cancel, but those in the line of motion survive and, in any given time interval, are

(6) $$\Delta p_n + \Delta p_m = 0,$$

(7) $$c\,\Delta p_n + \Delta \dot{p}_m = 0$$

where Δp_n (Δp_m) is the net change in momentum of the neutrino sea (mass) and ΔU_m is the kinetic energy gained by the mass. Eliminating Δp_n we get

(8) $$\Delta U_m = c \Delta p_m.$$

Since the kinetic energy is all supplied by the neutrino sea at velocity c = ds/dt, expression of (8) in differential form results in a force

(9) $$F \equiv dU_m/d_8 = dp_m/dt,$$

and Newton's second law is affirmed for matter interactions with the neutrino sea. Because the forces in (5) and (9) arise from the same basic scattering mechanism, the equivalence of inertial and gravitational mass is also affirmed.

It should be noted that the existence of a vast store of energy in the Universe which can interact with matter may be directly pertinent to the urgent need for nonchemical forms of energy suitable for human use.

NOTES

NOTES

NOTES

NOTES

NOTES

NOTES

NOTES

NOTES

NOTES

NOTES

NOTES

NOTES